RECENT PROGRESS IN SOLDERING MATERIALS

Edited by **Ahmad Azmin Mohamad**

Recent Progress in Soldering Materials
http://dx.doi.org/10.5772/66554
Edited by Ahmad Azmin Mohamad

Contributors

Ionelia Voiculescu, Ahmad Azmin Mohamad, Roman Koleňák, Wayne Chee Weng Ng, Sang Hoon Kim, Sangsun Yang

Published by InTech
Janeza Trdine 9, 51000 Rijeka, Croatia

Publishing Process Manager Nina Kalinic
Technical Editor SPi Global
Cover InTech Design team

First published November, 2017
Printed in Croatia
Legal deposit, Croatia: National and University Library in Zagreb

Additional hard copies can be obtained from orders@intechopen.com

Recent Progress in Soldering Materials, Edited by Ahmad Azmin Mohamad
p. cm.
Print ISBN 978-953-51-3643-9
Online ISBN 978-953-51-3644-6

For EU product safety concerns:
IN TECH d.o.o., Prolaz Marije Krucifikse Kozulić 3, 51000 Rijeka, Croatia,
info@intechopen.com or visit our website at intechopen.com.

Meet the editor

Dr. Ahmad Azmin Mohamad is an associate professor of Materials Engineering Program at the School of Materials and Mineral Resources Engineering, USM. He has 12 years of experience in the fields of corrosion and batteries. He is the author of over 50 publications and more than RM 3 million research grants.

Contents

Preface

In this progressive era of mobile technology, a demand for innovative electronic devices is flourishing and nonstop. These devices thrive on multitasking and excellent performances, which can be attributed to the interconnectivity of their components. Thus, the importance of solder materials in this industry is undoubtedly higher than before. Soldering has evolved from its previous function of joining metal pieces with a fusible metal alloy to broader aspects and requirements in its applications. The wide lead-free solder characterization, properties, mechanism, and applications of the latest developments in solder materials are reviewed in this book. Discussion topics are based on the latest new composition of alloy, updated characterization, and findings of solder joints, mechanical, and corrosion. For a full review on these topics, the reader is referred to all chapters in this book.

Ahmad Azmin Mohamad
Lecturer in Materials Engineering Program
School of Materials and Mineral Resources Engineering
Universiti Sains Malaysia
Nibong Tebal
Penang, Malaysia

Chapter 1

Introductory Chapter: Overview of Recent Progress in Soldering Materials

Ahmad Azmin Mohamad

Additional information is available at the end of the chapter

http://dx.doi.org/10.5772/intechopen.69764

1. Introduction

Sn-Pb solders have a long history and are widely used in the electronics industry due to their low cost, good solderability, low melting temperature and satisfactory mechanical properties. However, due to recent legislation and market pressures, Pb is being removed from electronic products [1].

Many lead-free solders have been studied as replacements for Sn-Pb solders. Of these, Sn-Ag-Cu-based materials are the most promising candidates owing to their overall properties, including mechanical and reflow properties, wettability and reliability.

Nowadays, electronic devices have become smaller and more complicated. However, the reliability of soldered components remains a critical issue. The need for high electric current density and the decreasing scale of solder-substrate interfaces of the latest advanced electronic products are everyday challenges for those in solder-related industries.

This book concerns solders and focuses on material characterizations, solder composition, and the methods used to make alloys and determine their structures, physical properties and applications. Physical properties, the factors that control them, and theoretical verification are key elements of solder research and will be reviewed in detail. Corrosion of solders is included in the coverage of the properties related to solder composition and mechanical properties.

2. Composition

To improve Sn-based solders for specific applications, various alloying elements have been studied, e.g., Ag, Bi, Al, Cu, In, Sb, Zn and so on. The selection of alloying elements is generally used to tackle problems regarding to pure Sn, to control the formation of intermetallic compounds (IMC), to keep the melting point at eutectic or near eutectic, to improve corrosion resistance and improve mechanical properties and so on [1, 2].

Several publications have already reported on these elements used to form various alloy compositions. Examples of basic compositions are Au-Sn, Bi-Sn, Sn-Ag, Sn-Cu, Sn-In and Sn-Zn. However, the properties of these binary alloy systems are far away from any applications [3]. In the next development, ternary solders are proposed, such as Sn-Ag-Cu, Sn-Ag-Bi and Sn-Zn-Bi solders. However, the formation of brittle failure behavior of IMC during soldering of both phases remains a problem. Now, trending in solder research is composite solders.

3. Joints

The interconnectivity of device components is very important in terms of electrical and mechanical properties. Most of these components are connected by solder joints. Solder joint is a very interesting subject to study in detail, and many new problems arise as electronic devices change almost daily.

Within this joint, IMCs are an important issue and must receive proper attention. Formation of an IMC layer at the solder/substrate interface after reflowing used to indicate a good joint between the solder and the substrate.

However, IMCs continue to achieve higher working temperatures and longer times. Thick IMC negatively affect the long-term reliability of solder joints due to their brittle nature [4, 5]. Even worse, to accommodate microelectronic components, current smaller and thinner sizes of solder joints are needed. The formation of smaller joints indirectly means that the volume fraction of the formed IMC layer tends to increase.

To inhibit interfacial IMC growth, several types of substrate finishes have been developed. The most popular surface finish for Cu substrates for high-end electronic applications is electroless nickel electroless palladium immersion gold (ENEPIG). ENEPIG is a tri-layered structure consisting of a layer each of electroless Ni, electroless Pd and immersion Au. The electroless Ni layer serves as an efficient diffusion barrier between the solder and the Cu pad, which can effectively inhibit the growth of interfacial IMCs [6].

Another approach to reducing the growth of IMCs uses composite solders. Inert reinforcements i.e. TiO_2, Al_2O_3, CeO_2, Fe_2NiO_4, TiC, TiO_2, ZnO, ZrO_2 and a carbon base i.e. graphene and carbon-nanotube are among the popular materials to be used in composite solders. These inert particles (mostly of nanosize) are nonreacting with the molten solder during the reflow process, which helps to refine the IMC's structure and consequently improve the mechanical and other properties [7–9]. Composite solders have therefore attracted considerable attention. For a full review of nanocomposite solders, the reader is referred to Shen and Chan [7].

4. Nanoindentation

Many physical/mechanical characterizations of solder joint have been proposed. However, since the miniaturization of devices reduces solder joint size, the effectiveness of characterization also needs to follow this trend. Nanoindentation hardness testing, using nanometer-sized indents, is another popular tool that is accorded with smaller component properties. It is used to estimate the physical and mechanical properties of materials [6]. An example of nanoindentation testing involves curve marks and plotting of the force that is applied to solder materials (**Figure 1**) [9].

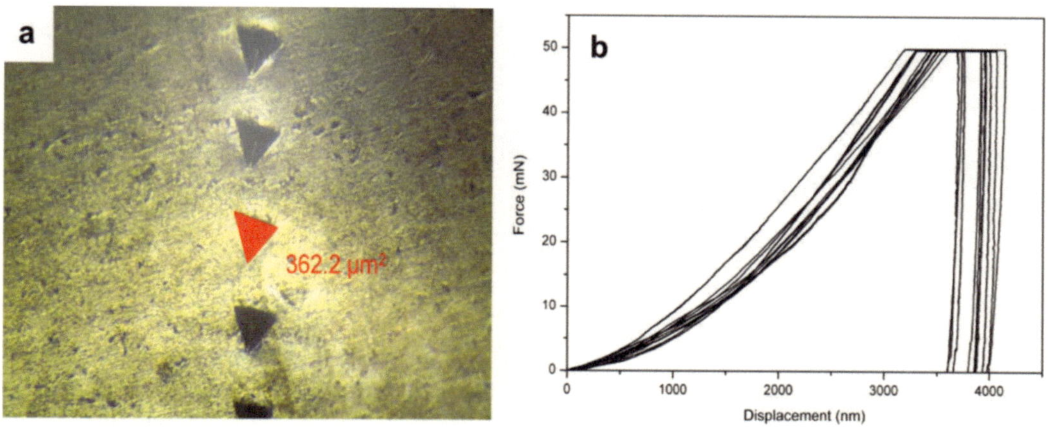

Figure 1. Areas of the nanoindentation marks on the cross-sectioned surface of (a) SAC305 and (b) the hysteresis plot of the loading and unloading indenters for SAC305, adapted from Ref. [9].

5. Corrosion

Material selections for solder are not normally designed for corrosion resistance. If this material is in contact with a corrosive environment, it could cause failure of the circuit. Once the corrosion has occurred, mechanical and electrical failures make the product stop. The corrosion properties of Sn-based lead-free alloys in corrosive environments have not been widely reported, even though it is important in many automotive, aerospace, maritime and defense applications [10]. Corrosion mainly originates either within the circuit (flux corrosion) or from the environment [11, 12].

5.1. Flux corrosion

Flux corrosion is related to the solder flux residue produced during soldering and remains corrosive even after the soldering operation has completed. Solder flux residue acts as a corrosion promoter in the presence of ionic substances and a resin component. Aggressive ions, such as Cl^- or Br^- in flux, will increase the corrosion activities. Flux residue resin also accumulates dust during operation and later provides a hydrophilic surface which forms a medium for the ions to react with the solder materials [12, 13].

The failure analysis of flux corrosion has been reported by several authors. Jellesen et al. [12], for instance, used a drop of solution (DI water or flux solution) on a micro tactile switch under DC bias. This reportedly helped to prove that the corrosion originated from flux contamination and condensation (**Figure 2**).

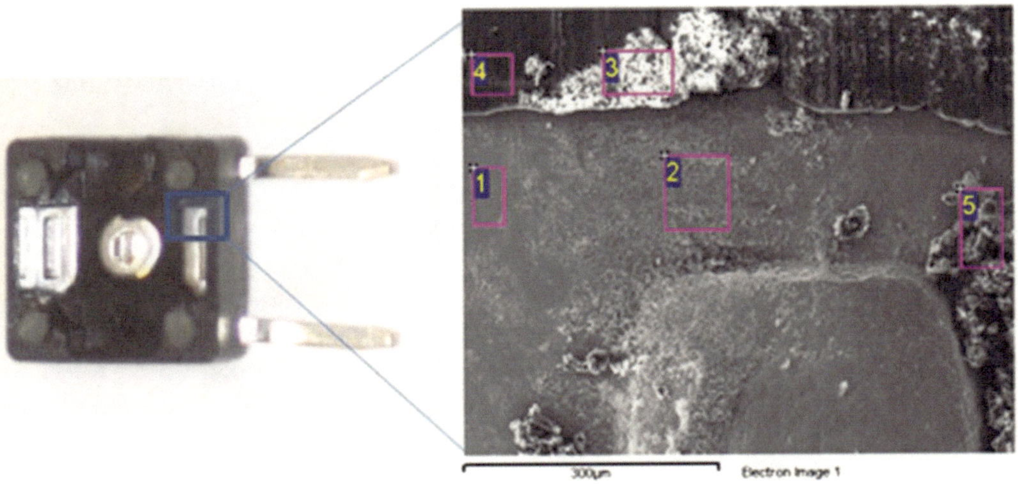

Figure 2. SEM micrograph of a failed switch due to severe corrosion and migration. (1) Silver, (2, 3) excessive deposits of Sn—major corrosion species, (4) carbon from the plastic housing and (5) flux residues, adapted from Ref. [12].

Figure 3. SEM image of a corroded Cu/Sn–9Zn/Cu butt joint after tensile strength measurement, adapted from Ref. [14].

5.2. Environment corrosion

Air moisture or aggressive mediums contained in natural environments are another source of solder corrosion. The most reported studies involved Cl^- and OH^- ions. Many alloy compositions and new elements have been introduced to improve corrosion resistance without reducing other properties.

Research on the corrosion of solders includes various aspects of solder characterizations. Examples include open circuit potential, galvanic cell, polarization, electrochemical impedance spectroscopy and so on.

One interesting corrosion study is the combined effect of corrosion on mechanical properties. Corrosion-mechanical studies are varied, e.g., the effect of immersion time of Cu/Sn-9Zn/Cu [14] (**Figure 3**). The focus here is more on the preferential dissolution of Zn and Sn. The formation of corrosion products and grooves proves the cause of joint failure. Later, the formation of cracks is the final stage that causes the mechanical properties of solders.

6. Conclusion

Lead-free solder characterization, properties, mechanism and applications of the latest developed solder materials are briefly reviewed in this introductory chapter. Further discussion on certain topics can also be found in this book. This review will hopefully assist readers in obtaining an overview of this exciting and promising field.

Author details

Ahmad Azmin Mohamad

Address all correspondence to: aam@usm.my

School of Materials and Mineral Resources Engineering, Universiti Sains Malaysia, Nibong Tebal, Pulau Pinang, Malaysia

References

[1] Kotadia HR, Howes PD, Mannan SH. A review: On the development of low melting temperature Pb-free solders. Microelectronics Reliability. 2014;**54**:1253-1273

[2] Affendy MG, Yahaya MZ, Mohamad AA. Corrosion of Sn-9Zn solder joints: A review. International Journal of Electroactive Materials. 2014;**2**:8-16

[3] Tsao L. Corrosion resistance of Pb-free and novel nano-composite solders in electronic packaging. In: Shin H, editor. Corrosion Resistance. INTECH Open Access Publisher; 2012. pp. 107-132

[4] Lee LM, Mohamad AA. Interfacial reaction of Sn-Ag-Cu lead-free solder alloy on Cu: A review. Advances in Materials Science and Engineering. 2013;**2013**:1-11

[5] Yang M, Ko Y-H, Bang J, Kim T-S, Lee C-W, Zhang S, Li M. Growth inhibition of interfacial intermetallic compounds by pre-coating oriented Cu6Sn5 grains on Cu substrates. Journal of Alloys and Compounds. 2017;**701**:533-541

[6] Ratzker M, Pearl A, Osterman M, Pecht M, Milad G. Review of capabilities of the ENEPIG surface finish. Journal of Electronic Materials. 2014;**43**:3885-3897

[7] Shen J, Chan YC. Research advances in nano-composite solders. Microelectronics Reliability. 2009;**49**:223-234

[8] Chen G, Huang B, Liu H, Chan Y, Tang Z. An investigation of microstructure and properties of $Sn_{3.0}Ag_{0.5}Cu\text{-}XAl_2O_3$ composite solder. Soldering & Surface Mount Technology. 2016;**28**:84-92

[9] Yahaya MZ, Ani FC, Samsudin Z, Sahin S, Abdullah MZ, Mohamad AA. Hardness profiles of $Sn\text{-}_{3.0}Ag\text{-}_{0.5}Cu\text{-}TiO_2$ composite solder by nanoindentation. Materials Science and Engineering: A. 2016;**669**:178-186

[10] Li D, Conway PP, Liu C. Corrosion characterization of tin-lead and lead free solders in 3.5 wt.% NaCl solution. Corrosion Science. 2008;**50**:995-1004

[11] Guédon-Gracia A, Frémont H, Plano B, Delétage J-Y, Weide-Zaage K. Effects of salt spray test on lead-free solder alloy. Microelectronics Reliability. 2016;**64**:242-247

[12] Jellesen MS, Minzari D, Rathinavelu U, Møller P, Ambat R. Corrosion failure due to flux residues in an electronic add-on device. Engineering Failure Analysis. 2010;**17**:1263-1272

[13] Smith BA, Turbini LJ. Characterizing the weak organic acids used in low solids fluxes. Journal of Electronic Materials. 1999;**28**:1299-1306

[14] Nazeri MFM, Mohamad AA. Effect of exposure to alkaline solution on Sn-9Zn solder joints. Journal of Materials Processing Technology. 2015;**219**:164-172

Chapter 2

Low Melting Temperature Solder Materials for Use in Flexible Microelectronic Packaging Applications

Sang Hoon Kim and Sangsun Yang

Additional information is available at the end of the chapter

http://dx.doi.org/10.5772/intechopen.70272

Abstract

The increasing application of heat-sensitive microelectronic components such as a multitude of transistors, polymer-based microchips, and so on, and flexible polymer substrates including polyethylene terephthalate (PET) and polyimide (PI), among others, for use in wearable devices has led to the development of more advanced, low melting temperature solders (<150°C) for interconnecting components in various applications. However, the current low melting temperature solders face several key challenges, which include more intermetallic compound formation (thus become more brittle), cost issues according to the addition of supplementary elements to decrease the melting point temperature, an increase in the possibility of thermal or popcorn cracking (reliability problems), and so on. Furthermore, the low melting temperature solders are still required to possess rapid electronic/electrical transport ability (high electrical conductivity and current density) and accompany strong mechanical strength sustaining the heavy-uploaded microelectronic devices on the plastic substrates, which are at least those of the conventional melting temperature solders (180–230°C). Thus, the pursuit of more advanced low melting temperature solders for interconnections is timely. This review is devoted to the research on three methods to improve the current properties (i.e., electrical and thermomechanical properties) of low melting temperature solders: (i) doping with a small amount of certain additives, (ii) alloying with a large amount of certain additives, and (iii) reinforcing with metal, carbon, or ceramic materials. In this review, we also summarize the overall recent progress in low melting temperature solders and present a critical overview of the basis of microscopic analysis with regard to grain size and solid solutions, electrical conductivity by supplementation with conductive additives, thermal behavior (melting point and melting range) according to surface oxidation and intermetallic compound formation, and various mechanical properties.

Keywords: heat-sensitive microelectronic components, flexible polymer substrates, low melting temperature solders (<150°C), interconnection applications

1. Introduction

More advanced solder bumps with low melting temperatures are crucial for use in flexible, bendable, and stretchable interconnection technology [1, 2]. In particular, the use of wearable devices requires the development of novel solders that can be reflowed at low temperature to avoid thermal damage to the usually temperature-sensitive components in these flexible devices, such as organic light-emitting diodes (OLEDs), polymer light-emitting diodes (PLEDs), and so on [3–6]. Thus, it is worthwhile to design a low melting temperature solder for more advanced interconnection technology and thus to impart more reliability to the solder bumps between future organic- or polymer-based microchips and flexible substrates. Furthermore, the continuous pursuit of multifunctionality in microelectronics has caused a significant increase in the number of solder bumps, to the level of 10,000 per chip [7]. It also has decreased the bump size to as small as 20 μm [7]. Thus, solder bumps with enhanced electrical and thermomechanical properties are needed to meet these demands. With these uses in mind, the properties of conventional solder materials with high melting temperatures (180–230°C) have been under scrutiny due to their reference characteristics, and whether implementing currently used soldering methods or inventing new ones, solution strategies to overcome problems associated with novel solder materials have been implemented.

In this chapter, we focus on the electrical and thermo-mechanical properties of novel solder materials with a specific range of low melting temperatures (<150°C). Emerging carbon reinforcement materials, such as carbon nanotubes (CNTs), graphenes, and their nanocomposites, are also briefly discussed and linked to the increasing development of composite solder materials (in addition to their low melting temperatures). In particular, strategies for improving the performance of solder materials are proposed, along with the provision of insight into classic metallurgy principles. To engineer the properties of low melting temperature solder materials in intended directions, new approaches using nanostructures, nanocomposites, alloying, and doping are also suggested.

2. Syntheses and/or fabrications of low melting temperature solders

In this section, we show how low melting temperature solders, consisting of various elements with higher melting temperatures, such as Sn (melting point of 231.9°C), In (melting point of 156.6°C), and Bi (melting point of 271.5°C), can be fabricated and/or synthesized using metallurgy principles or nanotechnology theory [8]. For example, the formation of an eutectic alloy with 42 wt.% Sn and 58 wt.% Bi can lead to a melting point decrease to 138°C due to the shift to the eutectic temperature [3, 4]. Due to the melting point drop according to the size decrease, the size reduction effect can also be used to synthesize a low melting point solder [4].

Low melting temperature Sn–In solder nanoparticles were successfully synthesized through a surfactant-assisted one-step chemical reduction method [9]. Different synthesis parameters, including pH, stirring speed, and surfactant concentration, were used to control the size, shape, and uniformity of the Sn–In solder nanoparticles [9]. At low In composition (20 wt.% In), the Sn–In solder nanoparticles were composed of $InSn_4$ alloy and β-Sn phase [9]. When the In content increased to

30 wt.%, the Sn–In solder nanoparticles were composed mainly of $InSn_4$, with a melting temperature of 115.5°C [9]. This low melting temperature indicates a new eutectic point for the Sn–In solder nanoparticle system, which is lower than that of the bulk alloy system (around 50 wt.% In) [9]. At higher In compositions, the Sn–In solder nanoparticles are composed of both $InSn_4$ and In phase [9].

To increase the compatibility and usefulness of the low melting point solder, eutectic Bi–43Sn nanowires with diameters of 20, 70, and 220 nm were fabricated by a hydraulic pressure injection process using anodic aluminum oxide templates [10]. Their microstructure investigation showed that the fabricated nanowires are composed of alternating Bi and Sn segments along the wire axes [10].

Novel Sn–Bi nanocomposites reinforced with reduced graphene oxide nanosheets (RGONs) were successfully fabricated by a simple, scalable, and economical electrochemical deposition method [11]. The Sn–Bi nanocomposites, reinforced with reduced graphene oxide nanosheets, had fine grain size as well as reduced graphene oxide nanosheets dispersed throughout the Sn–Bi matrix [11].

For the microstructural transformation and thermoelectrical improvement of Sn–Bi solder, MWCNT was reinforced using the electrochemical codeposition method [11]. Electron microscopy analysis can confirm that pristine MWCNTs were trapped in the deposited composites [11].

Bi-based solder powders with three chemical compositions (binary Bi–Sn, ternary Bi–Sn–In, and quaternary Bi–Sn–In–Ga alloy systems) were fabricated using a gas atomization technique; subsequently, the powders were further ball-milled to fabricate smaller-sized particulates; in particular, the diameter of those in the Bi–Sn–In–Ga powders became less than 10 μm with an irregular shape due to the nature of intrinsic brittleness, the degree of surface oxidation, and the formation of Ga intermetallic compound (IMC), all of which produced fractures of the Ga-containing powders [3].

Ternary Bi–Sn–In micropowders and nanoparticles were prepared as a composite solder material via gas atomization process and a chemical reduction method, respectively [3, 4]. To be specific, two types of Bi-based micropowders, composed of binary Bi–Sn and ternary Bi–Sn–In, were fabricated using a gas atomizer [4]. Then, the gas-atomized powders were classified using a series of standard sieves to obtain powders of a specific size range [4]. Bi (III) nitrate pentahydrate, Sn (II) 2-ethylhexanoate, In (III) nitrate hydrate at a Bi/Sn/In weight ratio of 43.5/31.5/25.0, and 1,10-phenanthroline were added to methanol, and the solution was stirred for 2 h [4]. Then, sodium borohydride was added, and the reaction continued at 50°C for 1 h [4]. The as-synthesized nanoparticles were centrifuged at 4000 rpm for 15 min, washed with methanol, and then dried in a vacuum oven for 24 h [4].

3. Classification of low melting temperature solders and terminologies for low melting temperature solder design

Conventional, low melting temperature solders are fabricated using either Sn–Bi or Sn–In [12]. In particular, Sn–Bi solders have received considerable attention because of their outstanding merits, including high wetting behavior, large creep resistance, and low coefficient of thermal expansion [12–14]. However, the relatively low mechanical strength and melting temperature (138°C) of these materials require improvement for their more effective use in flexible interconnection applications [3, 15]. Comparatively, Sn–In solder, which has a low melting temperature (118 °C), has excellent electrical and thermal properties [12, 16]. However, the price of In is very high, and this material includes high amounts of IMCs, which degrade the mechanical properties of the solder [16]. Thus, the incorporation of additives, including alloying, doping, or the use of reinforcement materials, into Sn–Bi or Sn–In alloy systems has been considered a useful strategy for improving the mechanical properties.

3.1. Microstructure with regard to grain size and solid solution

The microstructure reflects the mechanical properties of a solder [15–21]. Based on the microstructural analysis of low melting point solders, specific phases and their distribution in the microstructure can be observed, and these characteristics can be used to describe intended properties [15–21]. In this section, we show how the microstructure of low melting point solders is altered by the incorporation of an additive, especially with regard to grain size and solid solution.

The representative microstructure of eutectic Sn–Bi alloy is shown in **Figure 1**; in this structure, granular Sn-rich grains and similarly granular Bi-rich grains can be seen. The dark and bright gray regions are Sn and Bi, respectively; these regions appear as interlocked lamellar structures. Upon incorporation of an additive into the Sn–Bi solder, the solder microstructure was found to be remarkably transformed due to the formation of finer grains or the presence of new solid solutions or precipitates. Usually, a plausible explanation for this is that Sn- and Bi-rich grains are heterogeneously nucleated in the formation of certain IMCs. For example, Mokhtari et al. demonstrated that the addition of In or Ni can modify the microstructure of Sn–Bi solder; in particular, the addition of 0.5 wt.% In was able to suppress the coarsening of the Bi-rich phase, which means that the Sn–Bi–0.5In solder comprised primary Sn dendrites and eutectic phases [13, 15]. Comparatively, Ni appears to have been included in the Sn phase since Ni-containing Sn–Bi solder exhibited eutectic phases and the Ni_3Sn_4 IMC but did not show any sign of coarsening due to the Bi-rich phase.

Figure 2 exhibits the cross-sectional microstructure of a Sn–40Bi–2Zn–0.1Cu solder alloy composed of Sn-, Bi-, Zn-, and Cu-rich phases. For the cooling (solidification) process of the Sn–40Bi–2Zn–0.1Cu solder alloy from the liquid state, the following procedure took place: L (liquid) → L + primary Sn → primary Sn + eutectic (β-Sn + Bi-rich) + eutectic (β-Sn + Zn-rich) → primary Sn + secondary precipitated Bi + eutectic (β-Sn + secondary precipitated Bi + Bi-rich) + eutectic

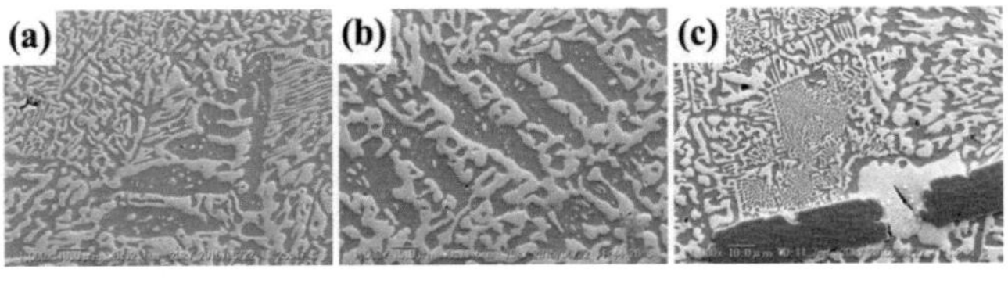

Figure 1. SEM micrographs of (a) eutectic Sn–Bi, (b) Sn–Bi–0.5In, and (c) Sn–Bi–0.5Ni [15].

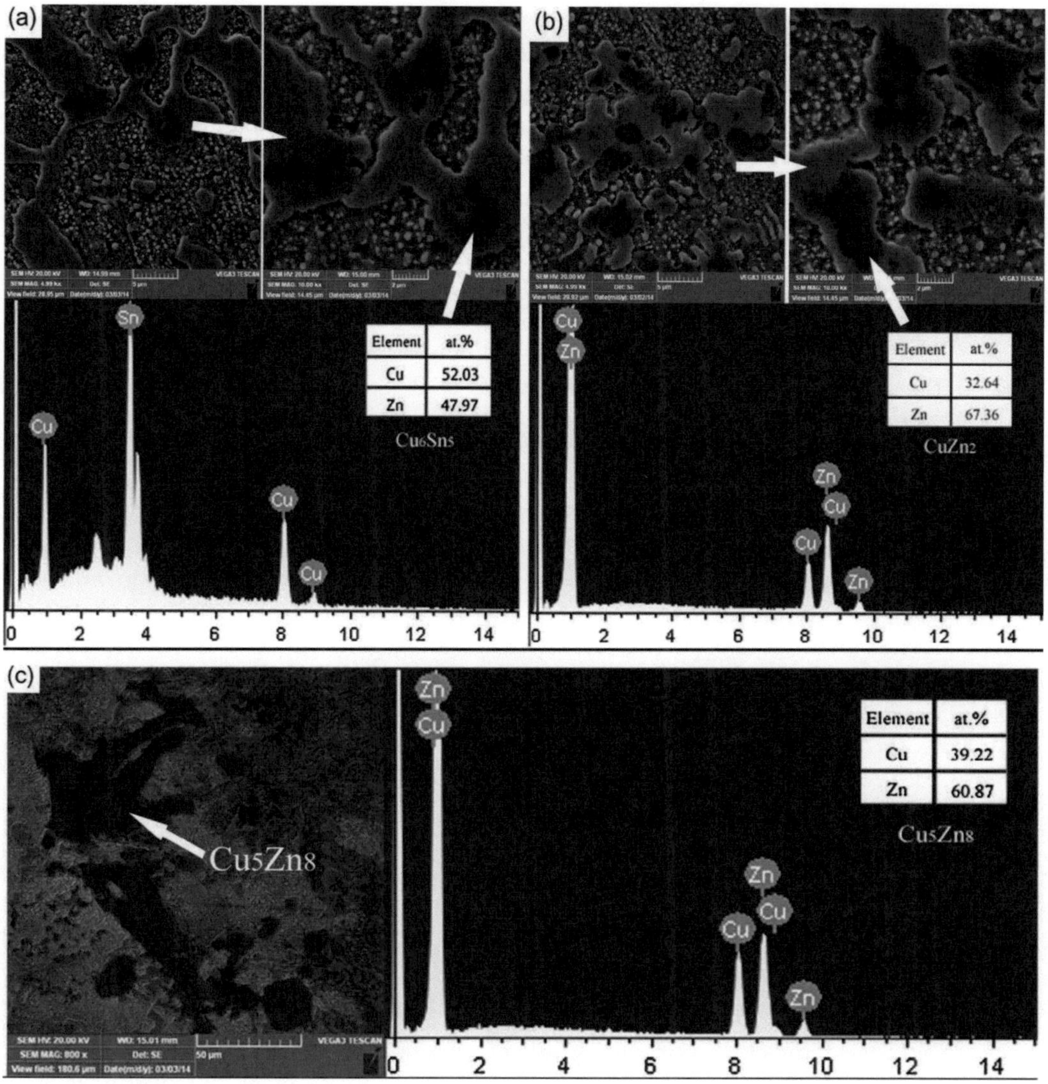

Figure 2. SEM micrograph and EDS analysis results of (a) Cu_6Sn_5 phase in Sn–40Bi–0.1Cu, (b) $CuZn_2$ phase in Sn–40Bi–2Zn–0.1Cu, and (c) Cu_5Zn_8 phase in Sn–40Bi–2Zn–0.1Cu [17].

(β-Sn + Zn-rich). Different from the reference Sn–40Bi–0.1Cu solder alloy, Cu_6Sn_5 precipitates were not found in the Sn–40Bi–2Zn–0.1Cu solder because Cu reacts more strongly with Zn than Sn does; thus, all of the Cu was consumed in the formation of Cu–Zn IMCs [17]. These results were also very similar to the studies of Islam and Li [22, 23]. According to the X-ray diffraction (XRD) and energy dispersive spectroscopy (EDS) analysis, globular $CuZn_2$ and blocky Cu_5Zn_8 precipitates were formed in Sn–40Bi–2Zn–0.1Cu solder; in particular, Zn atoms segregated between the Sn- and Bi-rich matrices and reacted with Cu atoms to form Cu–Zn IMC particles [17].

Figure 3 shows the microstructures of the Sn–58Bi and Sn–Bi–Sb alloys. The Sn–58Bi alloy shows a typical eutectic structure. The dark region is the Sn phase; the white region represents the Bi phase. For the Sn–Bi–Sb alloys, each part contains two phases, the dark and the light phases shown, respectively, in the images. When the Bi content changes, the proportion of each structure does not greatly change. However, the proportion of quasi-peritectic structure increases, as the Sb content increases [18]. For the Sn–Bi–Sb alloy, the intensity of the Sn-rich phases increased when the Sb content increased [18].

The near eutectic solder (In–50Sn) is presented in **Figure 4a**, showing a cross-sectional microstructure consisting of a combination of lamellar and irregular phases. In particular, the presence of a mixture of Sn-rich and In-rich phases and the In_3Sn IMC was determined by XRD and EDS analyses. Comparatively, **Figure 4b** shows the cross-sectional microstructure of the In–30Sn solder, which consisted of only two specific phases: an In-rich phase and the In_3Sn IMC but the absence of Sn-rich phases and/or other Sn IMCs [16].

As can be seen in **Figure 5**, the interfacial bonding of low melting temperature Sn–In solder on the Cu substrate is presented according to the reflow temperature and duration. The reflow at 180°C for 20 min selected was the optimal condition due to the high bonding strength of 6.5 MPa and the interfacial layer with less microvoids or mechanical cracks [20]. Especially, the reflow temperature of 180°C was high enough for the active diffusion of the low melting

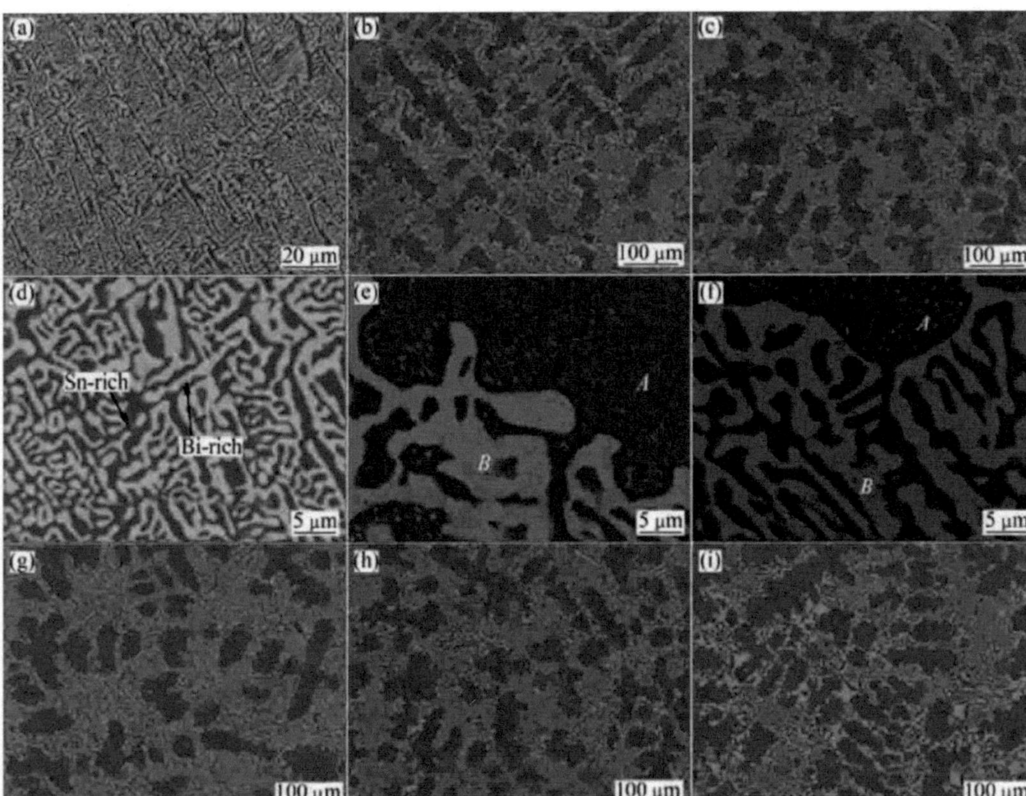

Figure 3. SEM micrographs of Sn–58Bi and Sn–Bi–Sb alloys: (a and d) Sn–58Bi, (b and e) Sn–52Bi–1.8Sb, (c and f) Sn–48Bi–1.8Sb, (g) Sn–48Bi–1.4Sb, (h) Sn–48Bi–1.8Sb, and (i) Sn–48Bi–2.4Sb [18].

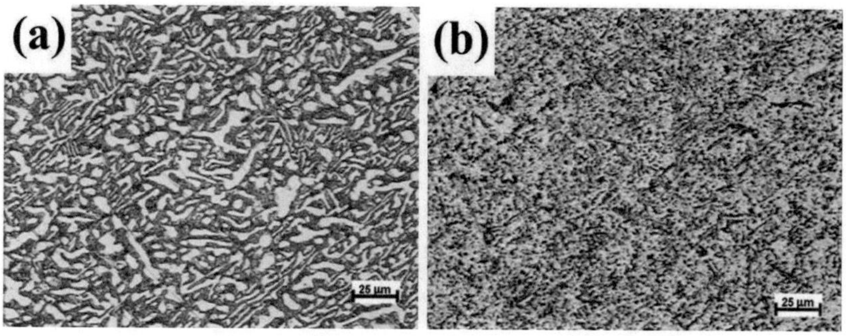

Figure 4. SEM micrographs of (a) In–50Sn and (b) In–30Sn solders [16].

temperature solders, although they were dissolved in the Cu substrates and formed IMCs such as $Cu_6(Sn, In)_5$ and $Cu_{11}(In, Sn)_9$ [20]. Therefore, the joint can sustain high service temperatures because it is formed completely from these IMCs.

Chen et al. determined the interfacial reactions in Sn–51In, Sn–20In, and Sn–20In–2.8Ag on Ag substrates reacted at various temperatures, with results shown in **Figure 6** [21]. Particularly for the Sn–51In/Ag couples, the reaction products are $AgIn_2$ and Ag_2In phases at 150°C and 100°C; only Ag_2In is formed at lower temperatures [21]. Due to the formation of different reaction phases, the reaction layer in the Sn–51In/Ag couples grows more slowly at 100°C than is the case for samples reacted at lower temperatures [21]. The interfacial reaction rates in the Sn–20In/Ag couples are much slower than those in the Sn–51In/Ag couples [21]. In the Sn–20In/Ag couples, the ζ-phase is formed at 250°C, and both ζ-Ag/$AgIn_2$ phases are formed at 125°C; however, no noticeable interfacial reactions are observed to have reacted at 75 and 100°C over a period of 1440 h [21].

The species of IMCs at the interface between the eutectic Sn–In solder and the single crystalline Cu substrate were systematically investigated using scanning electron microscopy (SEM) (**Figure 7**). After reflowing at 160°C for 5 s, two kinds of IMC were formed in three sublayers from the solder to the substrate side; the formed materials were a $Cu(In, Sn)_2$ layer with tetragonal crystal structure, a coarse-grain $Cu_2(In, Sn)$ sublayer, and a fine-grain $Cu_2(In, Sn)$ sublayer with hexagonal crystal structure [24]. The morphology of the $Cu(In, Sn)_2$ grains is chunk type, the largest grain size. In the process of increased liquid soldering, this $Cu(In, Sn)_2$ layer is prone to spalling into the solder, leaving a duplex structure of $Cu_2(In, Sn)$ as the dominating IMC, which should be paid attention during phase identification [24]. The fine-grain $Cu_2(In, Sn)$ shows a granule-type morphology with the smallest grain size; this material distributes homogeneously on the entire Cu substrates [24]. However, coarse-grain $Cu_2(In, Sn)$ is substrate dependent and has an elongated morphology on single crystalline Cu surfaces [24].

The SEM micrographs (**Figure 8**) show Sn–58Bi solders doped with different weight fractions of graphene nanosheets (GNSs). In particular, **Figure 8a** shows the typical lamellar structure of the eutectic Sn–58Bi solder, in which the dark regions represent the Sn-rich phase, while the white regions represent the Bi-rich phase. Compared to the

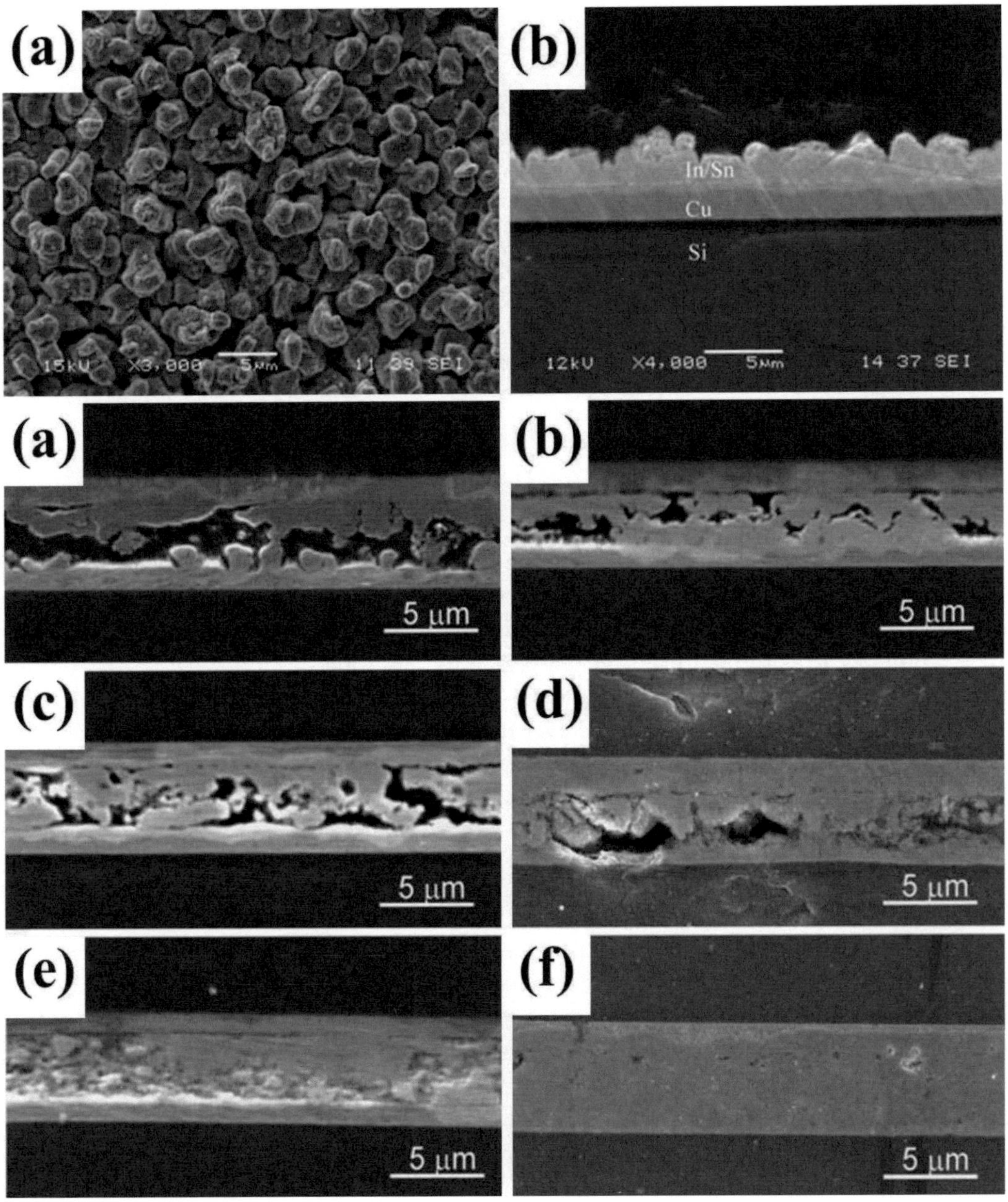

Figure 5. SEM micrographs (top) of as-received coating surface: (a) top view and (b) cross section. Cross-sectional SEM images (bottom) show that at the interfacial layers between the Sn-based solder and the cu substrate, the two components reacted and diffused under the following conditions (reflow temperature (°C) and duration (min)): (a) 140 and 5; (b) 140 and 20; (c) 160 and 5; (d) 160 and 20; (e) 180 and 5; and (f) 180 and 20 [20].

standard Sn–58Bi solder alloy, the microstructure of the composite solder is refined due to the increased content of graphene nanosheets, as shown in **Figure 8b–e**. The average grain size of the pure sample was about 1.6 μm, which is higher than that of the graphene

http://dx.doi.org/10.5772/intechopen.70272

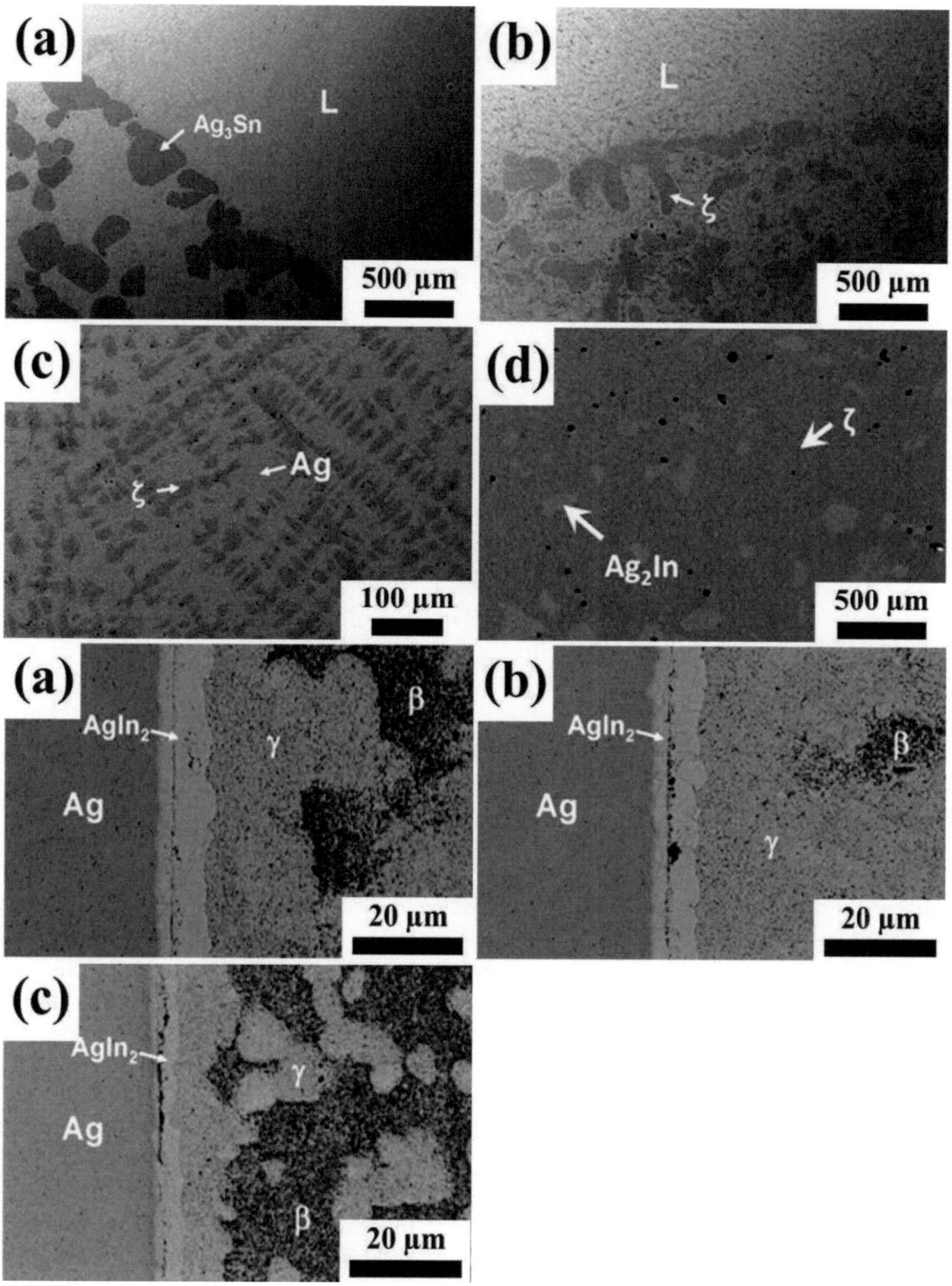

Figure 6. Backscattered electron imaging (BEI) micrographs (top) of (a) Sn–1.98 at.% In–8.03 at.% Ag, (b) Sn–10.02 at.% In–10.01 at.% Ag, (c) Sn–16.99 at.% In–80.01 at.% Ag, and (d) Sn–29.99 at.% In–68.99 at.% Ag annealed at 250°C for 12 weeks. BEI micrographs (bottom) of Sn–51In/Ag couple reacted at (a) 75°C, (b) 50°C, and (c) 25°C for 120 h [21].

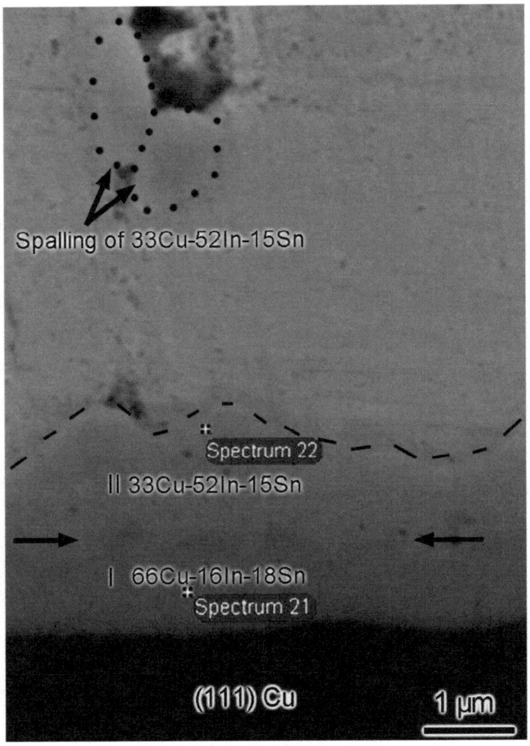

Figure 7. Cross-sectional SEM micrograph at the interface between eutectic Sn–In solder and single crystalline (1 1 1) Cu after reflowing at 160°C for 5 s [24].

nanosheet-doped solder samples [19]. The size reductions of the Sn–58Bi–graphene nanosheet (0.01, 0.03, 0.05, 0.1 wt.%) solders were about 28, 55, 30, and 32%, respectively [19]. These results indicate that the growth of grains was suppressed, and the microstructure of the Sn–58Bi solder alloys was refined by the addition of graphene nanosheets; in addition, the solder alloy with 0.03 wt.% graphene nanosheet addition showed the smallest average grain size [19]. The microstructure of local refinement can be found in **Figure 8b** for the 0.01 wt.% graphene nanosheet addition; meanwhile, uniform refinement of the structure and small Bi grains appear in the Sn–58Bi reinforced with 0.03 wt.% graphene nanosheets, shown in **Figure 8c**. These images indicate that the growth of grains can be suppressed by the presence of graphene nanosheets, which is caused by the high barrier of graphene nanosheets against the diffusion of metal atoms [19]. However, with more graphene nanosheets introduced, the growth of metal grains along the surface of the graphene nanosheets can be promoted [19]. As a result, the distinct feature of the uniformly distributed dendritic structure in Sn–58Bi reinforced with 0.05 wt.% graphene nanosheets can be seen in **Figure 8d**; this structure shows that graphene nanosheets can promote the growth of Bi dendrites. Moreover, local grain aggregation recurs and partial grains growth follows an inverse pattern when the content of graphene nanosheets reaches 0.1 wt.%, as shown in **Figure 8e**. This result is attributed to the increasing addition of graphene nanosheets, which undergo effective bonding with solders [19].

http://dx.doi.org/10.5772/intechopen.70272

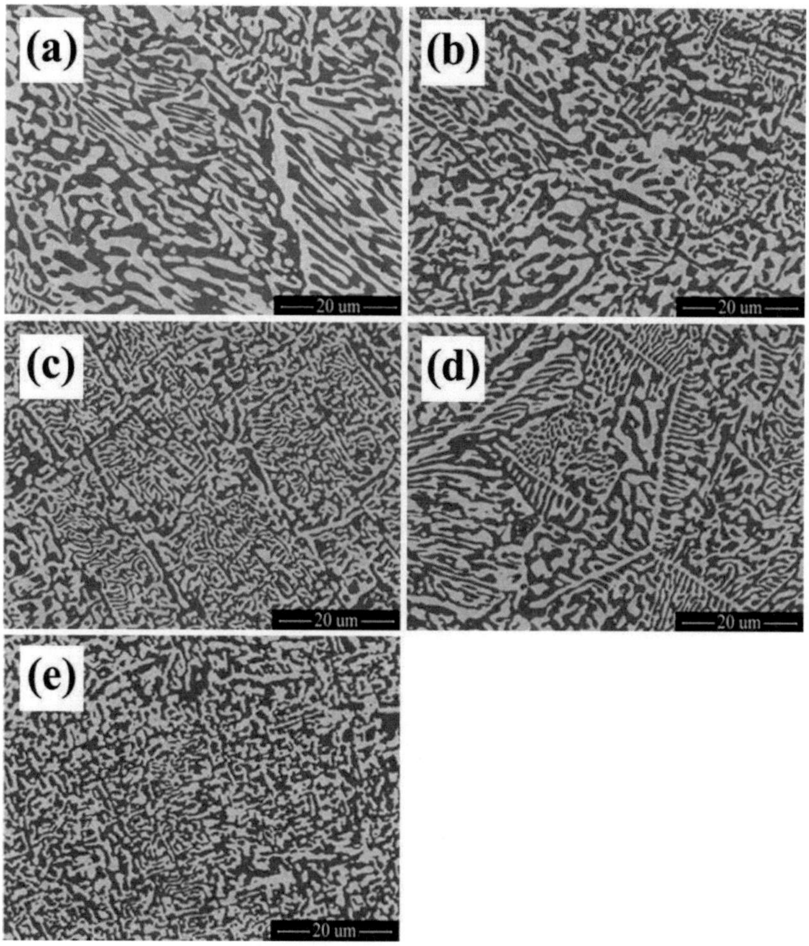

Figure 8. SEM micrographs of (a) Sn–58Bi, (b) Sn–58Bi–0.01GNSs, (c) Sn–58Bi–0.03GNSs, (d) Sn–58Bi–0.05GNSs, and (e) Sn–58Bi–0.1GNSs [19].

3.2. Electrical conductivity modification by supplementation with conductive additives cd, Sb, cu, Ag, Zn, in, Ni, and carbon nanomaterials

The solder bump size on a packaging substrate decreases as a result of the electronic components being miniaturized [25–28]. Simultaneously, the pitch distance drops to the submicron level [27, 28]. The pin count increases to meet the demands of rapid signal transmission and high current load [28]. Thus, the electrical property of a solder becomes one of the most important factors [29–31]. In the literature, however, there is not much research on the electrical properties of low melting point solders. Thus, determining the electrical property for a low melting point solder can be of great use to researchers and engineers, especially those who design solder alloys. Meanwhile, the eutectic Sn–Bi solder has a relatively high electrical resistivity of 30–35 μΩ·cm due to the electrical resistivity of Bi (115 μΩ·cm), while the eutectic Sn–In solder has a very low electrical resistivity of 10–15 μΩ·cm due to the electrical resistivity of In (8 μΩ·cm) [30].

Altıntas et al. determined that the electrical conductivity varies with temperature for low melting point solders: Sn–41.39 at.% Cd–6.69 at.% Sb, Sn–49 at.% In–1 at.% Cu, Sn–50 at.% Ag–10 at.% Bi, and Sn–32 at.% Bi–3 at.% Zn alloys; these values were determined by the four-point probe method, as shown in **Figure 9** [31]. The electrical conductivities of all solder alloys in the present work were found to decrease linearly with increasing temperature. The electrical conductivity values as a function of temperature were found to be in ranges of 4.35–2.76, 5.00–3.43, 5.30–4.58, and 1.52–1.39 (× 10^6)/Ω·m for Sn–Cd–Sb, Sn–In–Cu, Sn–Ag–Bi, and Sn–Bi–Zn solder alloys, respectively [31]. By extrapolating the electrical conductivity lines to their melting temperature, values of electrical conductivity for Sn–Cd–Sb, Sn–In–Cu, Sn–Ag–Bi, and Sn–Bi–Zn at their melting temperatures were determined to be 2.61, 3.26, 4.57, and 1.35 (× 10^6)/Ω·m, respectively, as shown in **Table 1**.

Figure 10 depicts the resistivity of all Sn–Bi solder alloys according to the increase of the Ni amount. No significant change in the electrical resistivity was detected following the addition of Ni. Although the existence of IMCs in the solders could induce a higher electrical resistivity, in this research the effect on the solders' electrical resistivity in the presence of Ni_3Sn_4 is not apparent [29].

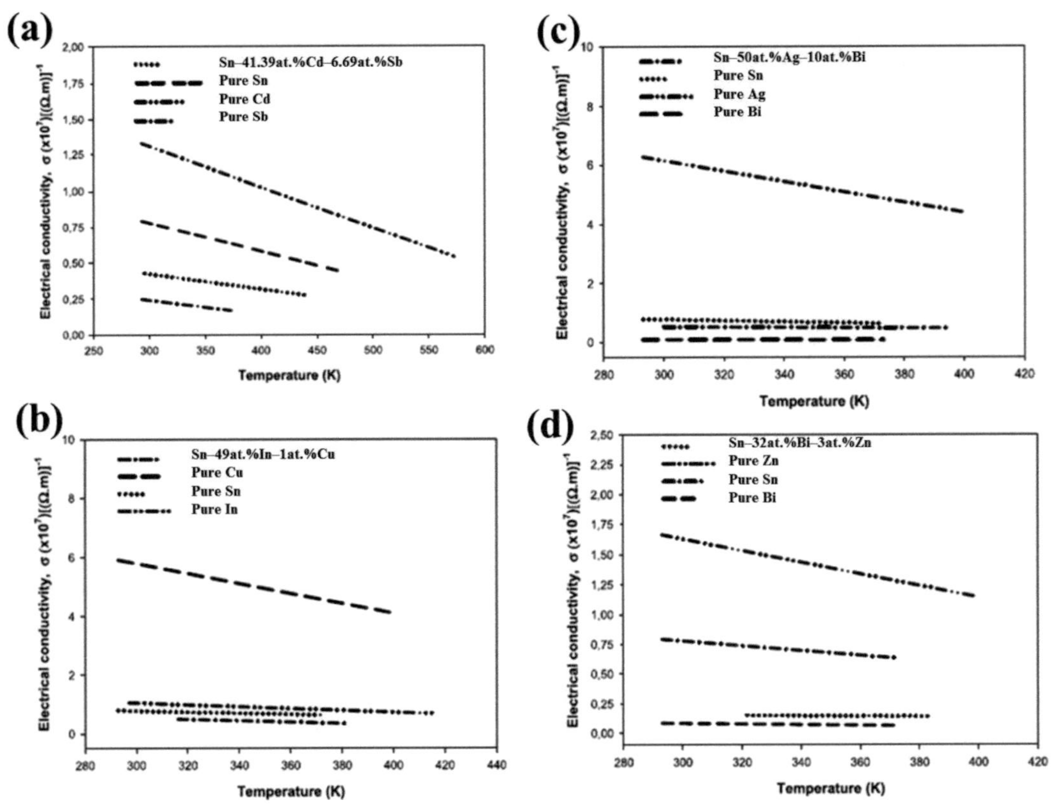

Figure 9. Electrical conductivity measurements according to temperature for (a) Sn–41.39Cd–6.69Sb, (b) Sn–49In–1Cu, (c) Sn–50Ag–10Bi, and (d) Sn–32Bi–3Zn (at.%) solder alloys [31].

	Melting temperature (K)	Temperature coefficient of σ (K^{-1}) × 10^{-3}	σ at the melting temp (1/Ω·m) × 10^6
Sn–41.39 at.% Cd–6.69 at.% Sb	453	2.47	2.61
Sn–49 at.% In–1 at.% Cu	389	4.97	3.26
Sn–50 at.% Ag–10 at.% Bi	411.4	1.14	4.57
Sn–32 at.% Bi–3 at.% Zn	404.7	1.00	1.35
Sn–4 wt.% Ag–2 wt.% In	490.7	–	–
Sn–20 wt.% Ag–2 wt.% In	490.7	–	–
Sn–40 wt.% Ag–2 wt.% In	490.7	–	–
Sn–20 wt.% In–25 wt.% Ag	490.7	–	–
Sn–20 wt.% In–10 wt.% Ag	490.7	–	–
Sn–20 wt.% In–15 wt.% Ag	486.0	–	–
Sn–6 wt.% Sb–5 wt.% Ag	507.8	–	–
Sn–42.8 wt.% Bi–0.04 wt.% Cu	411.8	–	–
Sn–3.5 wt.% Ag–0.9 wt.% Cu	490.2	–	–

Table 1. Some electrical properties of solid phase for Sn–41.39 at.% Cd–6.69 at.% Sb, Sn–49 at.% In–1 at.% Cu, Sn–50 at.% Ag–10 at.% Bi, Sn–32 at.% Bi–3 at.% Zn [31].

The incorporation of carbon nanomaterials with graphene structures can also impart much more rapid electron transfer than that can be obtained using conventional Sn–Bi solder [32]. Subsequently, reinforcement with carbon nanomaterials having high thermal conductivity can be used to tailor a network structure to effectively transfer the outer thermal energy to the solder matrix [32]. **Figure 11** shows the calculated current efficiency versus different MWCNT additions in the solder alloy. This implies that the current efficiency is dependent on the concentration of

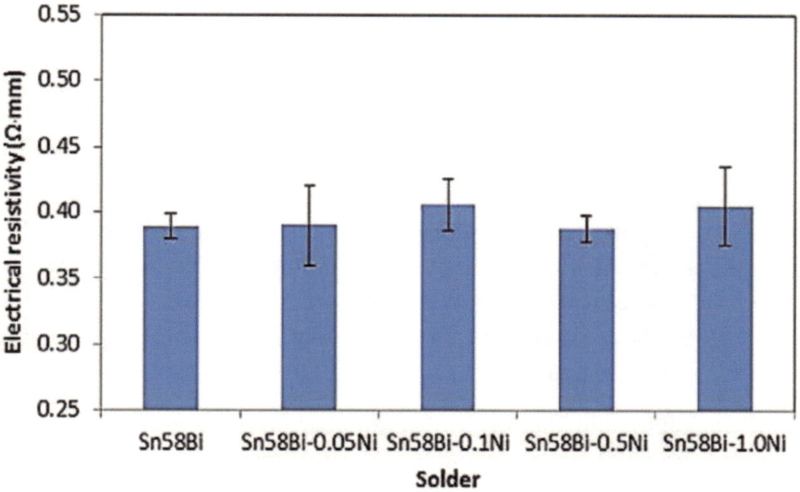

Figure 10. Electrical resistivity of Sn–58Bi–xNi [29].

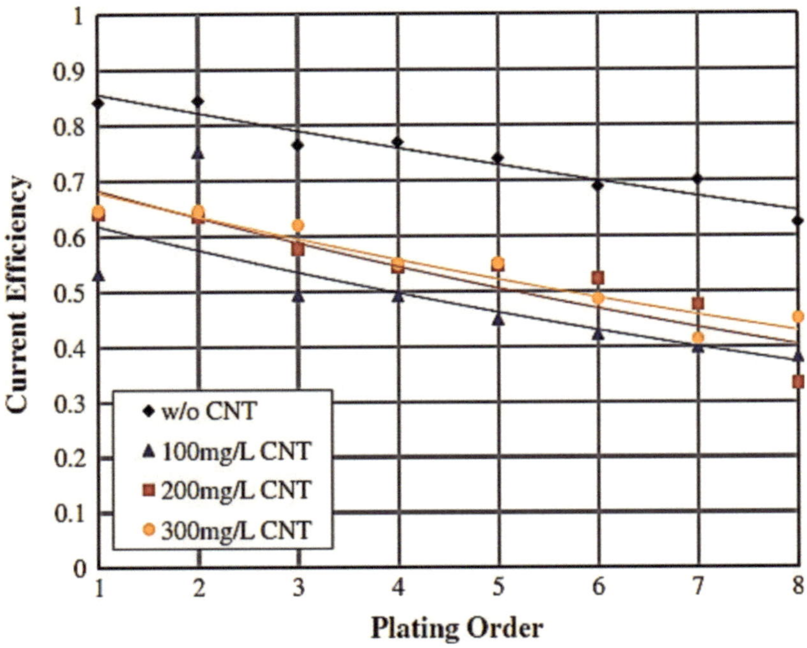

Figure 11. Current efficiency change of the Sn–Bi solder under CNT load [32].

the solder constituent electrolytes [32]. This is mainly due to the increased hydrogen evolution caused by lower ion concentration in the vicinity of the nucleation/deposition sites [32]. On the other hand, the addition of MWCNTs into the solder alloy reduces the current efficiency because pristine MWCNTs are trapped in the deposited composites [32]. Due to the bridging effects of trapped MWCNTs, the Sn–Bi–CNT composite is denser than the pure Sn–Bi alloy [32].

3.3. Thermal behavior (melting point and melting range) according to surface oxidation and intermetallic compound formation

Near-future generations of electronics are expected to be flexible, bendable, and wearable [1]. The use of flexible devices requires the development of a novel solder that can be reflowed at a low temperature to avoid thermal damage to these flexible devices, which usually have temperature-sensitive components [3, 4]. In addition, the melting point of a solder alloy should be the first priority for consideration when it comes to the manufacturing process [3, 4]. Meanwhile, the eutectic point (usually, the low melting point) in a binary phase diagram is where a liquid phase and two solid phases can coexist at equilibrium. Thus, a large number of low melting point solders with eutectic compositions are mostly used for flip chip solder joint applications between microchips and substrates.

Figure 12 shows the solidus and liquidus temperatures, and the differences between both temperatures, referred to as the melting range of Sn–58Bi–xNi (x = 0.05, 0.1, 0.5 and 1.0 wt.%) solder alloys. The addition of Ni apparently lowered the solidus and liquidus temperatures [29]. This phenomenon was attributable to the fact that the addition of small amounts of Ni altered the composition of the alloy to resemble the eutectic composition of the ternary Sn–Bi–Ni alloy system [29].

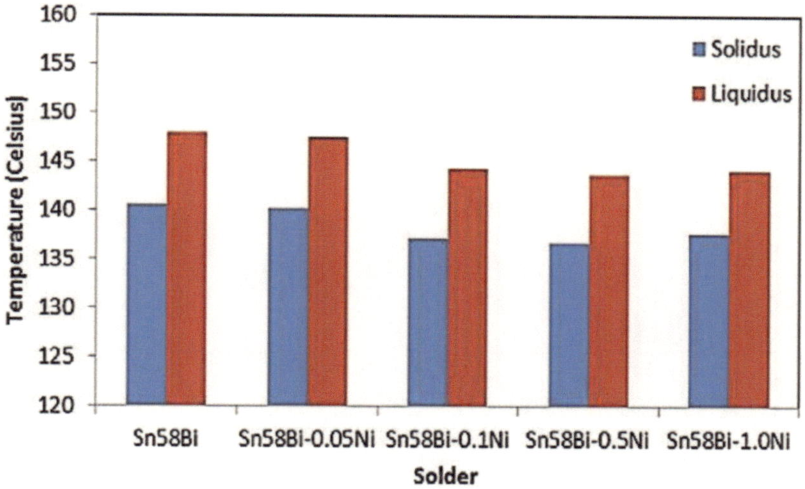

Figure 12. Solidus and liquidus temperatures of Sn–58Bi–xNi [29].

Using the differential scanning calorimetry (DSC), when the content of In was increased to 23.8 wt.%. Kim et al. determined that the prominent endothermic peak of the Bi–Sn–In powders shifted to 82.0°C from that of the Bi–Sn powders, which have a peak at 139.6°C [3, 4]. The continuous addition of 4.8 wt.% Ga shifted the peak even more to 60.3°C. Meanwhile, there was a slight broadening in the solidus line of the melting peak of the Bi–Sn–In solder powders mainly due to the formation of an In-rich phase [3]. Furthermore, the formation of new Ga0.9In0.1, BiIn, and In0.2Sn0.8 IMCs according to the addition of 4.8 wt.% Ga to the Bi–Sn–In solder alloy system also influenced the melting range broadening [3]. Kim et al. also show that ternary Bi–Sn–In nanoparticles, with a 71.1°C melting temperature, entered among the intervals of the higher melting temperature (79.4°C) micropowders and then reflowed at 110°C on a flexible polyethylene terephthalate (PET) substrate [4].

The fundamental thermal properties of Sn–58Bi–xZn (x = 0 and 0.7 wt.%) solder alloys were also analyzed by DSC, with results as shown in **Figure 13**. The results indicate that the solidus temperature of solder alloys slightly decreased with the addition of Zn content [33]. The reduction in solidus temperature of solders can probably be attributed to the increase in the surface instability due to the higher surface energy induced by the addition of Zn [33].

As can be seen in **Figure 14a**, the eutectic solder (In–Sn) had a low melting point of 118.5°C and a narrow melting range. The DSC curves of the hypo-eutectic Sn–70In and eutectic Sn–Bi solders were also presented in **Figure 14b** and **c**, respectively. In addition, the Bi53–Sn26–Cd21 solder presented in **Figure 14d** had the lowest melting temperature and a narrow melting range; thus, this solder had more active phase transformation than the others. The solidus temperatures, liquidus temperatures, and mushy temperature zones of Sn–58Bi, Sn–40Bi–0.1Cu, and Sn–40Bi–2Zn–0.1Cu solder alloys are collected in **Table 2**. The melting peak of the eutectic Sn–Bi solder decreased from 139.0 to 132.2°C according to the addition of a small amount of Cu; however, the addition of 2 wt.% Zn into the Cu-containing solder imparted a slight increase in the melting point (136.3°C) of

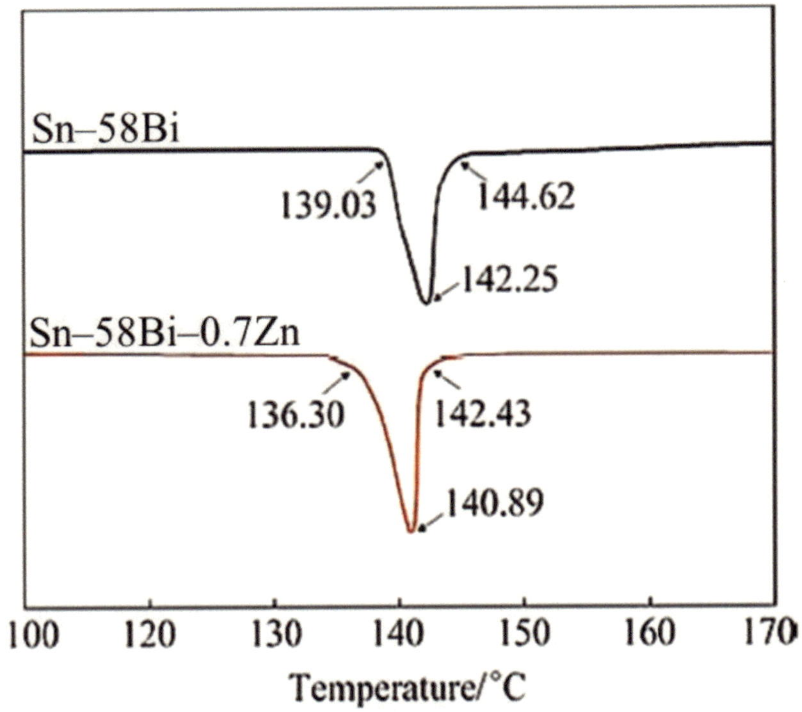

Figure 13. DSC curves of Sn–58Bi–xZn (x = 0 and 0.7) solders [33].

the Sn–40Bi–2Zn–0.1Cu solder [17]. Thus, the addition of Cu decreased the melting point of the Sn–Bi-based solder, while the addition of Zn provided the reverse effect (melting temperature increase) [17]. Moreover, Cu addition decreased the melting range of the Sn–Bi-based solder from 27.2 to 22.0°C, while the addition of Zn to Sn–40Bi–2Zn–0.1Cu increased the melting range slightly to 23.1°C [17]. The thermal conductivity of the Sn–40Bi–2Zn–0.1Cu solder of 24.51 W/(m·K) was the highest, while the Sn–40Bi–0.1Cu solder took second place with a value of 20.48 W/(m·K). Zn and Cu additions obviously improved the thermal conductivity of the Sn–Bi-based solder alloy [17]. Aksoz et al. reported that the thermal conductivity of pure Zn is 116 W/(m·K), which is higher than those of pure Bi (8 W/(m·K)) and pure Sn (67 W/(m·K)) [34]. This is the reason that the Sn–40Bi–2Zn–0.1Cu solder has the highest thermal conductivity [34]. The temperatures of the endothermic peaks of the nine Sn–Bi–Sb alloys are shown in **Table 3**. All the main peaks appear at around 147°C. The melting range of all the Sn–Bi–Sb alloys is larger than that of the eutectic alloy [18]. Side peaks are observed in many DSC profiles of the Sn–Bi–Sb alloys [18]. As the Bi content is reduced, the melting range obviously becomes large. Meanwhile, the melting range and the liquidus temperature reached maximum values for the composition of Sn–48Bi–1.8Sb and then started to drop when Sb content changed [18]. The melting range may be attributed to the fact that the proportion of the eutectic structure will change when Bi or Sb content changes [18]. For the liquidus temperature, it was found that the primary phase changes to the β-Sn phase, when the Sb content is more than 1.8% [18]. The presence of second phase implies that the remaining primary phase continues to melt after quasi-peritectic reaction [18].

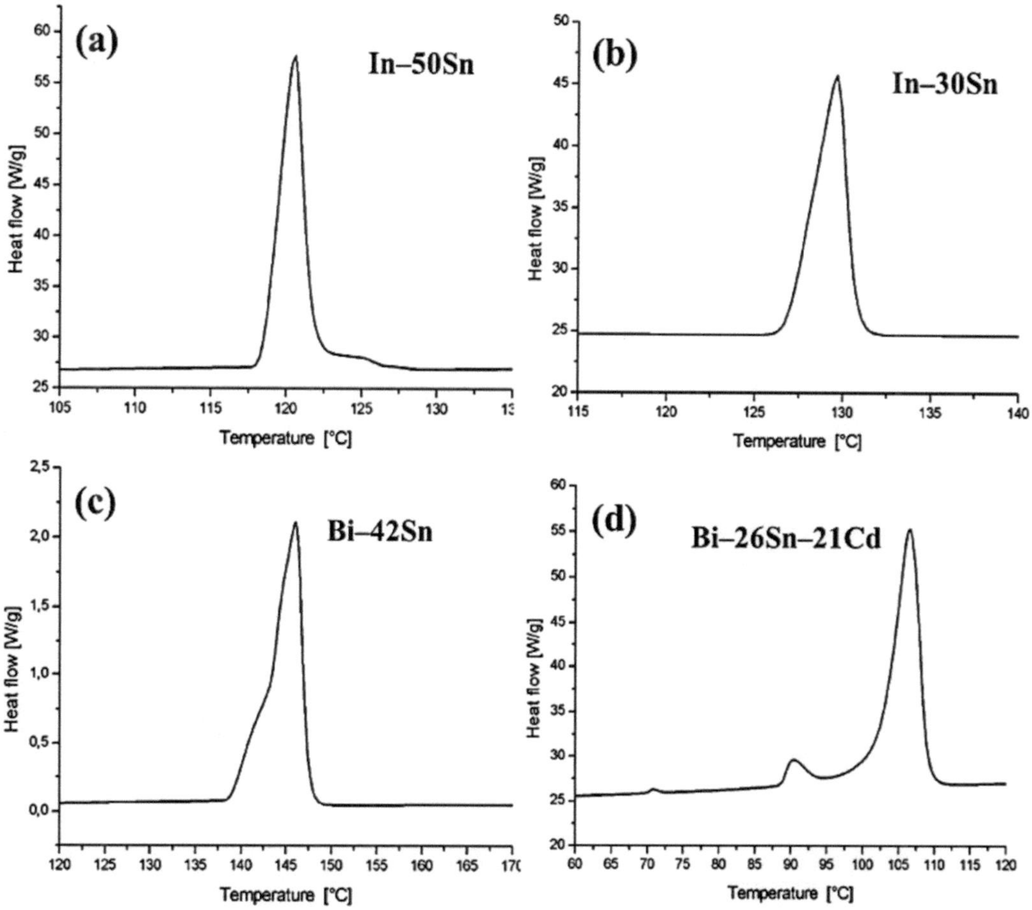

Figure 14. DSC curves of (a) In–50Sn, (b) In–30Sn, (c) Bi–42Sn, and (d) Bi–26Sn–21Cd solders [16].

Composition (wt.%)	Solid temperature (°C)	Liquid temperature (°C)	Pasty range (°C)	Mean TC (W/(mK))
Sn-58Bi	130.2	157.4	27.2	17.43 ± 0.6
Sn-40Bi-0.1Cu	125.1	147.1	22	20.48 ± 0.3
Sn-40Bi-2Zn-0.1Cu	127.7	150.8	23.1	24.51 ± 0.8

Table 2. Solidus temperatures, liquids temperatures, melting ranges, and mean thermal conductivity of the solder alloys [17].

Figure 15 shows the variation of the melting point of composite solder alloys with different dopant contents (both CNTs and Ni–CNTs) compared to the melting point of the Sn–57.6Bi–0.4Ag solder (about 140°C). It can be found that all the melting points were within the range of 139.3–139.6°C [35]. It has been reported that both CNTs and Ni–CNTs have an effect of reducing the melting point of solder alloys [35]. In particular, with the combined effect of CNTs and Ni–CNTs doped Sn–57.6Bi–0.4Ag solder alloys showed relatively lower melting points than those of CNTs doped solder alloys [35]. However, such small additions of CNTs and Ni–CNTs cannot have a significant influence on the melting point of Sn–57.6Bi–0.4Ag solder alloy [35].

Composition (wt.%)	Main peak temperature (°C)	Sec peak temperature (°C)	Solid temperature (°C)	Liquid temperature (°C)	Melting range (°C)
Sn–58Bi	143.1		139.4	148.0	8.6
Sn–52Bi–1.8Sb	147.7		140.6	152.0	11.4
Sn–48Bi–1.8Sb	146.5	163.0	140.9	172.7	31.8
Sn–44Bi–1.8Sb	146.9	169.0	141.9	180.5	38.6
Sn–48Bi–1.0Sb	144.7	162.0	140.6	168.7	28.1
Sn–48Bi–1.4Sb	146.8	163.3	141.2	170.4	29.2
Sn–48Bi–1.8Sb	146.5	163.0	140.9	172.7	31.8
Sn–48Bi–2.0Sb	147.6	164.4	142.3	169.7	27.4
Sn–48Bi–2.4Sb	148.5	163.3	142.8	169.3	26.5
Sn–48Bi–2.8Sb	148.0	162.6	143.6	168.4	24.8

Table 3. Thermal behaviors of standard Sn–58Bi alloy, Sn–52Bi–1.8Sb alloys, and Sn–48Bi–xSn alloys [18].

Figure 16 shows the DSC endothermic peaks of Sn–Bi nanocomposites reinforced with 0.02 or 0.05 g of reduced graphene nanosheets. While a large endothermic peak corresponding to the melting reaction in the range of 139.0°C of Sn–Bi solder has been observed, it was found that the melting point of Sn–Bi nanocomposites reinforced with reduced graphene nanosheets was about 139.0°C, which indicates that there was no significant effect on the thermal behavior of the nanocomposite solder, despite of the addition of reduced graphene nanosheets [11].

3.4. Mechanical properties

The durability and reliability of electronic products, as related to the mechanical properties of the solder joints, have become very important [13, 36–41]. This is especially true for portable, wearable devices, which frequently experience mechanical shock loadings caused by external forces [4]. Particularly for drop tests, during which the strain rate is very high, high mechanical shock resistance of solders is needed for these materials to fulfill their roles of structural materials. In addition, low melting point solders experience significantly high stresses during the reflow process owing to thermal gradient difference [3, 4]. Thus, there has been continuous interest in better understanding of the mechanical properties and in inventing high durability and reliability low melting point solders. One frequently utilized way to influence the mechanical properties of low melting point solder joints in a given system is to either alloy the materials or add small or large amounts of additional elements. In particular, any metal oxides or impurities may have marked effects on the mechanical properties of low melting point solders. Additional elements can fundamentally influence the mechanical properties of low melting point solders. First, additional elements can have an influence on the mechanical properties of the interfacial reactions between the solder and the substrate. Second, additives can positively change the mechanical properties of low melting point solders. Third, they can impart negative side effects, which result in a sacrifice of other mechanical properties of low

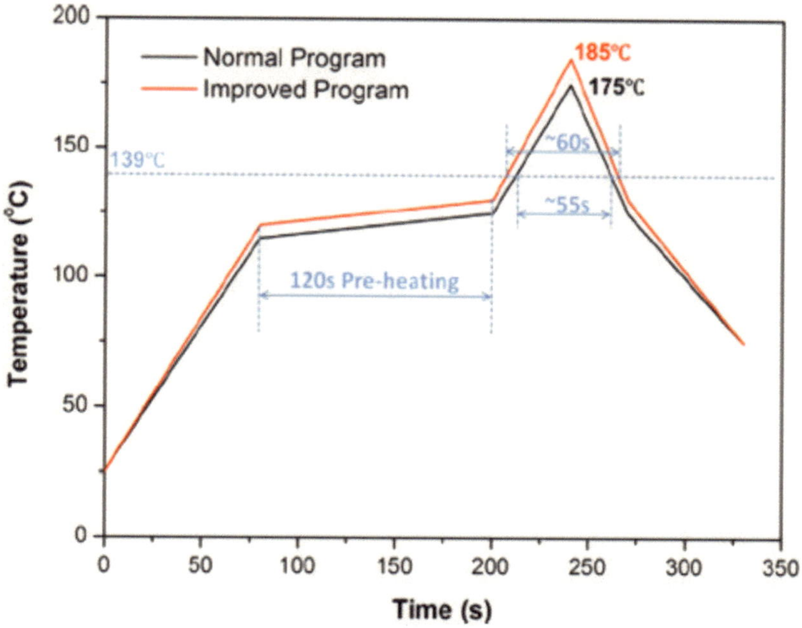

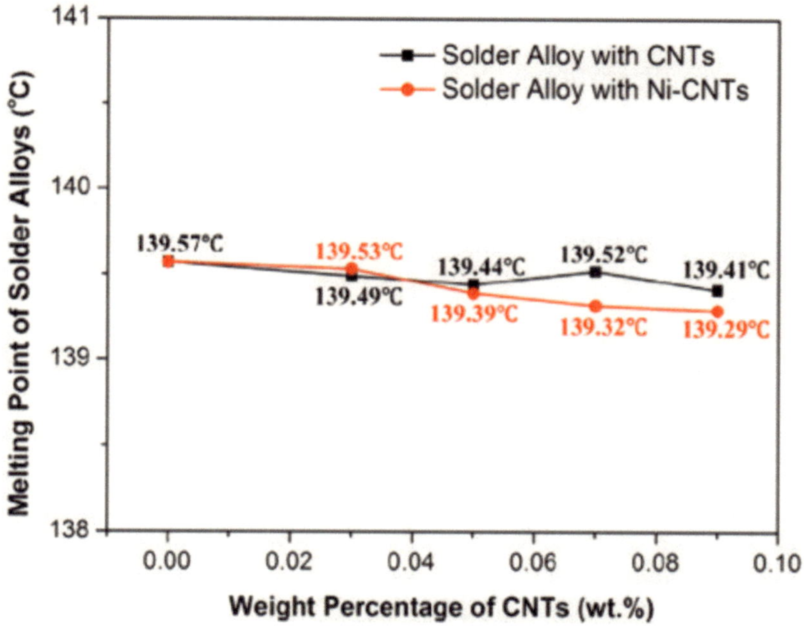

Figure 15. Plot of the variation of melting point of solder alloys with different amounts of CNTs or NI–CNTs [35].

melting point solders. In this section, therefore, we report on a number of investigations about the effects of different alloying elements, as well as the effects of metal oxides or impurities, in low melting point solders.

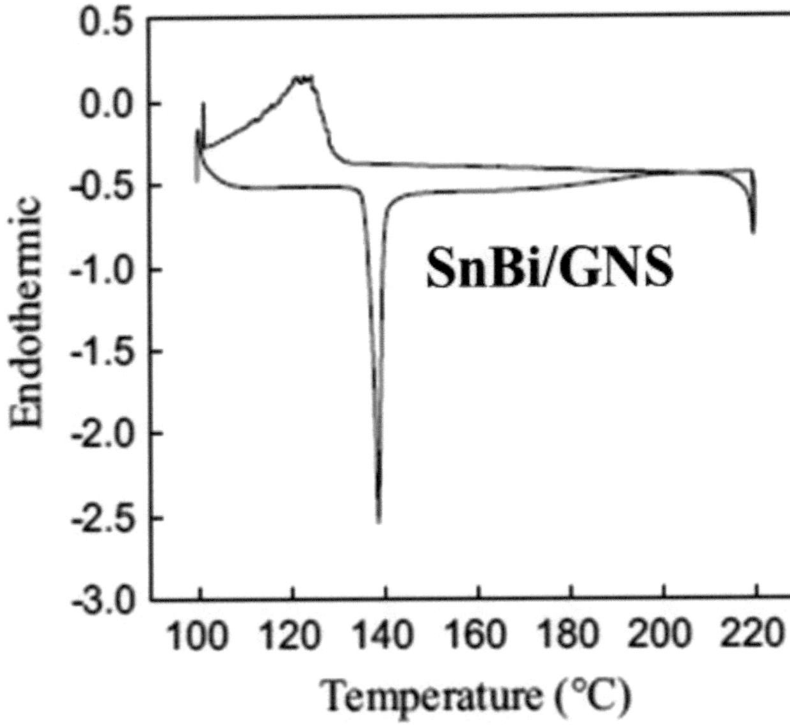

Figure 16. DSC of Sn–bi/RGOS nanocomposites with Sn content of 36.0 wt.% [11].

After the addition of 0.05 wt.% Cu_6Sn_5 nanoparticles in Sn–Bi solder, the tensile properties of the solder underwent brittleness caused by a change in ductility [36]. However, nanoindentation testing revealed that the creep resistance of the Sn–Bi–Cu_6Sn_5 solder is enhanced through the creep mechanism transformation [36]. In corrosion experiments, samples with Cu_6Sn_5 nanoparticles exhibit a lower corrosion rate [36].

Adding different sized Ag nanoparticles to a eutectic Sn–Bi alloy system refined the grain (microstructure), suppressed the growth and expansion of the interfacial IMCs, and increased the shear strength of the solder joint [38]. To be specific, the reinforcement with 76 nm Ag nanoparticles refined the microstructure by 49.1% and enhanced the microhardness by 12.2% compared to the standard Sn–Bi solder because the extent of the formation of the Cu–Sn IMC decreased from 0.394 to 0.339, suppressing the IMC thickness by 39.7% and improving the shear strength by 18.9% after reflowing at 220°C for 180 min. However, after the addition of both larger (133 nm) and smaller (31 nm) Ag nanoparticles, such thermomechanical properties improvements were lower than those of the solder having 76 nm Ag nanoparticles [38]. These improvements might have been due to refinement and dispersion strengthening and adsorption [38]. In fact, although the solder with smaller Ag nanoparticles should have had higher property improvements than those with larger ones, the agglomeration of the smaller sized Ag nanoparticles deteriorated the overall solder's properties and reduced the practical improvements [38]. Overall, an optimal particle size was proposed to balance the theoretical improvement and the agglomeration weakening; this size generated the best real improvement [38].

The reinforcement effects of the Al_2O_3 nanoparticles in Sn–58Bi solder were investigated from the aspects of electromigration, shear strength, and microhardness [39]. The experimental results show that the Al_2O_3 nanoparticles significantly improved the mechanical performances of the solder joints. The addition of Al_2O_3 nanoparticles reduced the thickness of the Bi IMCs along the interfacial layers [39]. More specifically, the growth rate of the IMC thickness according to the addition of Al_2O_3 nanoparticles decreased by 8% compared with that of pristine solder [39]. Furthermore, the microhardness of Al_2O_3-containing solder exhibited better performance than that of pristine solder according to aging time [39]. On the other hand, the addition of Al_2O_3 significantly improved the shear strength of the solder joint after aging for 48 and 288 h [39]. More specifically, after the solder was aged for these time periods at 85°C, the amplitudes of the shear strength increased by 3.5% and 2.4%, respectively, because unlike the smooth surface of the pristine solder, the surface of the Al_2O_3-containing solder showed a ductile failure (fractured) state [39].

To improve the mechanical behaviors of a Sn–58Bi/Cu joint, a minor amount of elemental Zn was alloyed into the Cu substrate [40]. The interfacial IMC growth and bending properties of Sn–58Bi/Cu and Sn–58Bi/Cu–2.29Zn were studied according to the effect of isothermal liquid and solid aging [40]. Although there was no significant change in the composition, thickness, or morphology of the interfacial IMC under liquid aging, the depressing of IMC growth at the interface between the Sn–58Bi solder and the substrate and the avoidance of the formation of Cu_3Sn IMC, Kirkendall voids, and Bi segregation at the IMC/Cu interface were realized for the Cu–Zn substrate under isothermal solid aging [40]. Joint strength and fracture behavior were also improved when using the Cu–Zn substrate [40]. There was no obvious decrease in the joint strength, and fracturing during bending was found mainly to occur in the solder matrix with ductile fracture mode or along the solder/IMC interface with partly brittle fracture mode for the Cu–Zn joint; these behaviors can be compared with the dramatically decreased joint strength and brittle fracture mode that occurred along the interface between IMC and Cu in Sn–Bi/Cu joints after aging [40].

Sn–57.6Bi–0.4Ag solder was reinforced with tungsten (W) nanoparticles at a concentration of 0.5 wt.% [41]. Due to the dispersion of W nanoparticles and the consequently refined microstructure, the mechanical properties of the solder alloy were enhanced, as indicated by the 6.2% improvement in the microhardness [41]. During electromigration, the segregation of the Sn-rich and Bi-rich phases and the accumulation of an (Au, Ni), $(Sn, Bi)_4$ layer at the cathode interface were also alleviated by the addition of W nanoparticles, which improved the electromigration resistance [41].

The tensile properties of eutectic Sn–Bi, Sn–Bi–0.5In, and Sn–Bi–0.5Ni solder alloys, and their shear strength as Cu/solder/Cu joints were investigated [15]. The addition of 0.5 wt.% Ni decreased the elongation property of the Sn–Bi alloy because of the formation of Ni_3Sn_4 IMCs [15]. The In-bearing solder alloys exhibited the greatest elongation among all the tensile-tested solder alloys [15]. The eutectic Sn–Bi and Ni-bearing solder joints exhibited degraded shear strength owing to the formation of coarsened Bi-rich phases [15]. The thermally aged Sn–Bi solder joints on Cu substrates exhibited a harshly fractured surface structure in the IMC layers at the interfacial boundaries, whereas the thermally aged In- and Ni-containing Sn–Bi solder joints showed a smoothly fractured surface structure because of the growth suppression of Cu–Sn IMCs [15]. In particular, the as-reflowed In-containing solder joints had a dimple-like, fractured surface structure, indicating a ductile microstructure because the thermally aged In-containing solder joints retained their ductile property well, while both the coarsened, fractured surface structure and excessive IMC growth of the thermally aged Sn–Bi solder joints at the interfacial boundaries is able to explain their mechanical degradation [15].

Four different concentrations of Ni (i.e. 0.05, 0.1, 0.5, and 1.0 wt.%) were individually added to Sn–58Bi samples, and respective microstructure, tensile strength, elongation, and wettability of Sn–58Bi–xNi were subsequently measured [29]. The results indicate that Ni refined the microstructure of the solder matrix and induced the formation of the Ni_3Sn_4 phase; furthermore, the formation and then continuously increasing concentration of Ni_3Sn_4 were proportional to the increase of Ni added to the solder [29]. Thus, the optimal concentration of Ni added to enhance the solder's tensile strength should be less than 0.1 wt.% [29]. Nevertheless, the elongation of the alloy was in fact inversely proportional to the increase of the added Ni content, although the appropriate incorporation of Ni contributed positively to the wettability of the solder alloy [29].

When the In content increased to 4% in the Sn–Bi alloy, tensile test results showed that the tensile strength increased slightly with the increase of added In, while the elongation first increased remarkably and then decreased after the addition of 2.5 wt.% In [13]. The diffused In was confirmed to participate in interfacial reactions, thereby forming Cu–Sn–In IMCs and affecting the wettability of the Sn–Bi solder on the Cu substrate [13]. Tensile strength changed slightly with increasing In addition, while the elongation increased remarkably with the addition of 2.5 wt.% In [13].

The interfacial reaction kinetics, tensile strength, and creep resistance of the Sn–58Bi–xZn (x = 0.0 or 0.7 wt.%) solder samples during liquid-state aging were investigated [33]. With the addition of 0.7 wt.% Zn, ultimate tensile strength (UTS) values of the eutectic Sn–Bi solder increased by 6.05 and 5.50% after soldering and aging, respectively; those values for the Cu/Sn–Bi/Cu solder joints also increased by 21.51 and 29.27%, respectively [33]. The increase in strengthening of the Cu/Sn–Bi–xZn/Cu solder joints can be attributed to the phase transformation at each Cu/IMC/solder interface due to the formation of finer Bi grains according to the addition of Zn [33].

The effect of Sb content on the mechanical properties of Sn–Bi solders was studied [18]. The mechanical properties of the solders/Cu joints were also evaluated [18]. The results show that the ternary alloy solders contain eutectic structures resulting from a quasi-peritetic reaction [18]. With the increase of the Sb content, the size of the eutectic structure increases [18]. A small amount of Sb has a large impact on the wettability of the Sn–Bi solders [18]. Reaction layers form during the spreading process [18]. Sb is detected in the reaction layer, while Bi is not detected [18]. The total thickness of the reaction layer between the solder and Cu increases with increased Sb [18]. The shear strength of the Sn–Bi–Sb solders also increases as the Sb content increases [18].

The mechanical properties of the melt-spun Bi–42Sn, Bi–40Sn–2In, Bi–40Sn–2Ag, and Bi–38Sn–2In–2Ag alloys were studied using dynamic resonance and Vickers indentation techniques at room temperature and compared to the mechanical properties of the traditional Sn–Pb eutectic alloy [42]. The results show that the crystallographic structure of the Bi–42Sn alloy presents as a combination of body centered tetragonal Sn and rhombohedral Bi [42]. The two ternary alloys exhibit additional constituent phases of $SnIn_{19}$ for Bi–40Sn–2In and Ag_3Sn for Bi–40Sn–2Ag alloys [42]. Attention has been paid to the role of IMCs in the mechanical and creep behavior [42]. The In- and Ag-containing solder alloys exhibited a good combination of higher creep resistance as compared with the Pb–Sn eutectic solder alloy [42]. This was attributed to the strengthening effect of Bi in the Sn matrix and the formation of $InSn_{19}$ and Ag_3Sn IMCs, which act as grain refiners in the matrix material [42].

Sn–57.6Bi–0.4Ag solder joints with different contents of CNTs and Ni–CNTs were investigated [35]. In particular, it was possible to improve the mechanical properties of the Sn57.6Bi0.4Ag solder joints by the addition of either CNTs or Ni–CNTs, and those with the addition of 0.05 wt.% CNTs or 0.07 wt.% Ni–CNTs showed the best mechanical performance [35]. With the addition of either CNTs or Ni–CNTs, the solder joints had rougher, fractured surface structures, resulting in better bonding properties. Although reinforcement with either CNTs or Ni–CNTs improved the mechanical performance of solder joints, Ni–CNTs worked much better [35]. Ni coating was proven to significantly inhibit the aggregation of CNTs, which can induce cracks and wetting problems and even deteriorate the strength of solder joints [35].

Sn–Bi composite solders containing Ni–CNTs were successfully synthesized [43]. The mechanical properties of Sn–Bi with different weight percentages of Ni–CNT were investigated [43]. The UTS and elongation of the Sn–Bi–0.05(Ni–CNT) solder with the optimized amount (0.05 wt.%) of Ni–CNTs increased remarkably because the CNTs and Ni_3Sn_4 enhanced the wettability and bondability of the composite solder [43]. However, because of the presence of CNT clusters and the intrinsic brittleness of IMCs, the UTS and elongation degraded with increased addition of Ni–CNTs [43]. That is to say, the UTS of the solder joint reached its maximum value with 0.05 wt.% Ni–CNTs addition and then degraded after increased addition of Ni–CNT [43]. Moreover, the tensile strength of the composite solder was much higher than that of the pristine solder, and subsequently, the creep resistance and hardness of the Sn–Bi–0.05(Ni–CNT) solder increased significantly compared to those of the Sn–Bi solder [43]. However, the hardness and the creep performance also decreased with 0.1 and 0.2 wt.% CNT content due to the same reasons mentioned above [43]. The CNT clusters and pore formation in the presence of the IMC with its intrinsic brittleness contributed to the decreases of hardness and creep performance [43].

The mechanical strength and ductility of the eutectic Sn–Bi solder alloy were dependent on the incorporation of MWCNTs [44]. Mechanical test results show that the bending strength of the Sn–Bi–0.03CNT composite increased by 10.5% compared to that of the reference Sn–Bi alloy, which can be attributed to the reduction in Sn-rich segregation and to grain refinement [44]. In particular, the toughness of the Sn–58Bi–0.03CNT composite increased by 48.9% compared to that of the unreinforced Sn–Bi solder alloy [44]. In addition, corresponding fracture surface comparison between the Sn–58Bi–0.03CNT composite and the monolithic Sn–58Bi alloy was performed to identify the influence of CNTs on the fracture behavior [44].

The effects of graphene nanosheets on the mechanical properties of the Sn–58Bi–0.7Zn solder joint were investigated [45]. Experimental results and finite element simulations showed that the best mechanical property improvement came from the 0.076 wt.% graphene nanosheet-doped Sn–58Bi–0.7Zn sample [45]. For the thermal aging samples, the UTS of the solder joint was also increased by 2.04% [45].

The mechanical properties (the stress expansion and strain distribution during a single lap shear test) of the Sn–Bi–graphene nanocomposite according to the weight ratio of graphene were simulated based on the theoretical calculations of the finite element method [46]. The strength of the joint was found to be mainly influenced by the shear stress; initial cracking was found to occur at the edge of the joint [46]. The shear modulus of the Sn–Bi–graphene nanocomposite was 192% greater than that of the pure Sn–Bi alloy, when the content of graphene increased to 1.0 wt.% [46]. Stress concentration was found to exist near the edge of the graphene, where initial failure may occur [46].

Graphene nanosheets were successfully incorporated at various percentages (0, 0.01, 0.03, 0.05, or 0.1 wt.%) into Sn–58Bi solder [19]. The tensile properties, wettability, corrosion resistance, microhardness, and creep behavior were subsequently improved [19]. Tensile and nanoindentation tests reveal that the composite solder with 0.1 wt.% graphene nanosheets leads to enhancements of about 14 and 38%, respectively [19]. With 0.01 wt.% graphene nanosheet addition, the elongation is 49% greater than that of the pure Sn–58Bi solder alloy [19]. The creep performance and the corrosion resistance are all enhanced by addition of graphene nanosheets [19]. The mechanism of enhancement of the graphene nanosheets of the performance of the composite solder alloy is also analyzed in this work [19]. Tensile tests reveal that the UTS of the solders rises gradually with graphene nanosheet addition; there is a 14% enhancement of tensile strength for the Sn–58Bi–0.1graphene nanosheet [19]. The huge enhancement of 49% in the elongation of Sn–58Bi–0.01graphene nanosheet and the establishment of a brittle to ductile fracture mode are induced by the strengthening effect of graphene nanosheets [19]. The wettability is improved with graphene nanosheet addition because the nanosheets lower the interfacial surface energy between the solder and the substrate [19]. Moreover, the corrosion resistance is distinctly enhanced in the Sn–58Bi–0.1graphene nanosheet, and this material retains a lower corrosion rate than that of Sn–58Bi [19]. The hardness and creep resistance leads to an obvious improvement due to the addition of graphene nanosheets [19]. The hardness is enhanced by 38%, when the addition of graphene nanosheets increases to 0.1 wt.% [19]. The enhancement of the creep behavior is further illustrated by the variation of the creep mechanism in the solder alloys [19]. Among the composite solders synthesized, the Sn–58Bi–0.1graphene nanosheet provides the best tensile strength and hardness with decreased ductility (**Table 4**) [19].

Compositions	Additives	Causes	Results	References
Sn–58Bi	0.05 wt.% Cu_6Sn_5 nanoparticles	Brittle IMC nanoparticles	Creep resistance	[36]
Sn–58Bi	76 nm Ag nanoparticles	Refinement and dispersion	Microhardness and shear strength	[38]
Sn–58Bi	Al_2O_3 nanoparticles	Reinforcement effect	Electromigration resistance, shear strength, and microhardness	[39]
Sn–58Bi/Cu and Sn–58Bi/Cu–2.29Zn joints	Alloying 2.29 wt.% Zn to Cu substrate	Suppression of Cu_3Sn formation, Kirkendall voids, and Bi segregation at the IMC/Cu interface	Adhesion strength	[40]
Sn–57.6Bi–0.4Ag	0.5 wt.% W nanoparticles	Refined microstructure	Electromigration resistance and microhardness	[41]
Sn–58Bi, Sn–58Bi–0.5In, and Sn–58Bi–0.5Ni	0.5 wt.% In or Ni	Ni_3Sn_4 IMC formation at the grain boundaries	Shear strength degradation of Sn–58Bi–0.5Ni and improvement of Sn–58Bi–0.5In	[15]
Sn–58Bi	0.05, 0.1, 0.5, or 1.0 wt.% Ni	Ni_3Sn_4 IMC formation	Wetting behavior	[29]
Sn–58Bi/Cu joint	2.5 wt.% In addition to the solder	Cu–Sn–In IMCs	Wetting behavior	[13]
Sn–58Bi	0.0 or 0.7 wt.% Zn addition to the solder	Liquid-state aging	Ultimate tensile strength	[33]
Sn–58Bi/Cu joint	Sb addition	Increasing size of the grain	Wetting behavior	[18]
Sn–58Bi	2 wt.% In, 2 wt.% Ag, or each 2 wt.% In and Ag	$SnIn_{19}$ and Ag_3Sn IMCs (grain refiner)	Creep resistance	[42]
Sn–57.6Bi–0.4Ag	0.05 wt.% CNTs or Ni–CNTs	Ni_3Sn_4 IMCs	UTS and elongation	[43]
Sn–58Bi	0.03 wt.% MWCNTs	Reduction of Sn-rich segregation and refinement	Bending strength	[44]
Sn–58Bi–0.7Ag	Graphene nanosheets	Finite element simulation	UTS	[45]
Sn–58Bi	Graphene	Finite element method	Shear modulus	[46]
Sn–58Bi	0, 0.01, 0.03, 0.05, or 0.1 wt.% graphene nanosheets	Reinforcement effect	Tensile strength, wettability, corrosion resistance, hardness, and creep behavior	[19]

Table 4. The improvement of various mechanical properties of the low melting temperature solders.

4. Strategies for enhancing the electrical and thermo-mechanical performance of low melting temperature solder materials

The electrical and thermo-mechanical properties of many different low melting point solders are affected by a variety of processing factors, such as the reflow temperature, time, flux used and its effectiveness, and temperature of measurement [15, 36, 37, 40, 41, 45]. Furthermore, the change of chemical composition of solders according to the addition of supplementary additives is also considered as a main factor that modifies or even improves a solder's electrical and thermo-mechanical properties [15, 36, 37, 40, 41, 45]. Three main methods are used to improve the electrical and thermomechanical performance of low melting point solders: (i) doping with a small amount of certain elements via diffusion reactions, (ii) alloying with a large amount of certain elements, and (iii) reinforcing with metal or ceramic elements. Sometimes, (iv) all of these methods are combined for the enhancement of the intended properties of low melting temperature solders.

Transient liquid phase bonding was conducted using the Sn–Bi solder with 30 wt.% Cu particles added [37]. However, this process caused the melting point of the solder joints to increase from 139 to 201°C; In addition, the solder joints contained large voids, resulting in a considerable degradation in shear strength [37].

On a Cu substrate, the conventional Sn–58Bi solder was alloyed with 0.7 wt.% Zn to improve the interfacial reaction, tensile strength, and creep resistance during liquid-state aging [37]. However, the overgrown IMC layers between the Sn–58Bi solders and Cu substrates significantly degraded the reliability of the electronic products [37].

Lin et al. added minor amounts of Ga, ranging from 0.25 to 3.0 wt.% to Sn–58Bi solder [14]. As a result, the growth of IMC layers was effectively suppressed [14].

Hu et al. fabricated an Sn–58Bi composite solder reinforced with Al_2O_3 nanoparticles to slow down electromigration and to improve the shear strength and microhardness [39].

Four different concentrations of Ni (0.05, 0.1, 0.5, and 1.0 wt.%) were individually added to the Sn–58Bi solder [29]. The optimal concentration of Ni necessary to enhance the tensile strength of the alloy was 0.1 wt.%, but the elongation of the alloy was inversely correlated to the Ni content [29].

Wojewoda-Budka and a coworker demonstrated excellent diffusion soldering process results for Bi–22 at.% In on Cu interconnections; this was proved by the presence of $Cu_{11}In_9$ phase present in the Cu/In–22Bi/Cu interfaces in the temperature range of 85–200°C [47].

Graphene nanosheets were successfully incorporated at various percentages (0, 0.01, 0.03, 0.05, and 0.10 wt.%) into Sn–58Bi solder; the microstructure, tensile properties, wettability, corrosion resistance, hardness, and creep behavior were significantly improved [19].

Sun et al. introduced a low melting temperature Sn–57.6Bi–0.4Ag solder reinforced with different concentrations of MWCNTs or Ni-coated MWCNTs [35]. With the addition of

MWCNTs and Ni–MWCNTs, the fractured surface of the solder joints became rougher, leading to a better bonding structure [35]. Though both MWCNTs and Ni–MWCNTs have the capability to improve the mechanical performance of solder joints, Ni–MWCNTs worked much better [35]. The Ni coating was proved to significantly inhibit the aggregation of MWCNTs, which can solve cracks and wetting problems and even improve the strength of solder joints [35].

5. Concluding remarks

A recent trend in solder research mentioned that low melting temperature solder materials and their nanocomposite materials will be suitable for flexible interconnection applications in the near future. Thus, fabrications and/or syntheses, as well as elaboration of the electrical and thermomechanical properties, of various low melting temperature solder materials are discussed in detail. The various determination factors regarding the electrical and thermomechanical properties of solder materials are also elucidated with theoretical and experimental support. Subsequently, a promising approach to enhancing the performance of solder materials using supplementary additives, such as nanostructures, nanocomposites, alloying, and doping, is described with examples. It is possible to conclude that low melting temperature solders may enable significant advancement in interconnecting components in various applications and soldering technologies for the flexible microelectronic packaging industry.

Acknowledgements

This study was supported financially by Fundamental Research Program of the Korea Institute of Materials Science (KIMS). This work was also supported by the National Research Council of Science & Technology (NST) grant by the Korea government (MSIP) (No. CAP-12-6-KIMS).

Conflict of interest

With regard to competing financial interests, the authors declare that they have none.

Author details

Sang Hoon Kim and Sangsun Yang*

*Address all correspondence to: nanoyang@kims.re.kr

Powder Technology Department, Korea Institute of Materials Science, Changwon, Republic of Korea

References

[1] Gates BD. Materials science. Flexible electronics. Science. 2009;**323**:1566-1567

[2] Ko Y, Kim M, Bang J, Kim T, Lee C. Properties and reliability of solder microbump joints between Si chips and a flexible substrate. Journal of Electronic Materials. 2015;**44**:2458

[3] Kim SH, Son MJ, Nguyen VL, Lim T, Yang D, Kim M, Kim KB, Kim YJ, Lee JH, Kim YD. Preparation of property-controlled bi-based solder powders by a ball-milling process. Metals. 2016;**6**:74-84

[4] Kim SH, Yang D, Kim Y, Min T, Choi J, Yun J, Kim KB, Kim YJ, Lee JH, Do Kim Y. Thermomechanical evolution of ternary Bi–Sn–In solder micropowders and nanoparticles reflowed on a flexible PET substrate. Applied Surface Science. 2017;**415**:28-34

[5] Liu Y, Fu H, Sun F, Zhang H, Kong X, Xin T. Microstructure and mechanical properties of as-reflowed Sn–58Bi composite solder pastes. Journal of Materials Processing Technology. 2016;**238**:290-296

[6] Zardetto V, Brown TM, Reale A, Di Carlo A. Substrates for flexible electronics: A practical investigation on the electrical, film flexibility, optical, temperature, and solvent resistance properties. Journal of Polymer Science Part B: Polymer Physics. 2011;**49**:638-648

[7] Ho C, Lee P, Chen C, Yang C. Electromigration in 3D-IC scale Cu/Sn/Cu solder joints. Journal of Alloys and Compounds. 2016;**676**:361-368

[8] Frongia F, Pilloni M, Scano A, Ardu A, Cannas C, Musinu A, Borzone G, Delsante S, Novakovic R, Ennas G. Synthesis and melting behaviour of Bi, Sn and Sn–Bi nanostructured alloy. Journal of Alloys and Compounds. 2015;**623**:7-14

[9] Shu Y, Rajathurai K, Gao F, Cui Q, Gu Z. Synthesis and thermal properties of low melting temperature tin/indium (Sn/In) lead-free nanosolders and their melting behavior in a vapor flux. Journal of Alloys and Compounds. 2015;**626**:391-400

[10] Chen S, Chen C, Luo Z, Chao C. Fabrication and characterization of eutectic bismuth–tin (Bi–Sn) nanowires. Materials Letters. 2009;**63**:1165-1168

[11] Peng Y, Deng K. Fabrication of reduced graphene oxide nanosheets reinforced Sn–Bi nanocomposites by electro-chemical deposition. Composites Part A: Applied Science and Manufacturing. 2015;**73**:55-62

[12] Chang S, Wu S. Low-frequency damping properties of eutectic Sn–Bi and In–Sn solders. Scripta Materialia. 2011;**64**:757-760

[13] Chen X, Xue F, Zhou J, Yao Y. Effect of In on microstructure, thermodynamic characteristic and mechanical properties of Sn–Bi based lead-free solder. Journal of Alloys and Compounds. 2015;**633**:377-383

[14] Lin S, Nguyen TL, Wu S, Wang Y. Effective suppression of interfacial intermetallic compound growth between Sn–58wt.% Bi solders and Cu substrates by minor Ga addition. Journal of Alloys and Compounds. 2014;**586**:319-327

[15] Mokhtari O, Nishikawa H. Correlation between microstructure and mechanical properties of Sn–Bi–X solders. Materials Science and Engineering A. 2016;**651**:831-839

[16] Chriašteľová J, Ožvold M. Properties of solders with low melting point. Journal of Alloys and Compounds. 2008;**457**:323-328

[17] Shen J, Pu Y, Yin H, Luo D, Chen J. Effects of minor Cu and Zn additions on the thermal, microstructure and tensile properties of Sn–Bi-based solder alloys. Journal of Alloys and Compounds. 2014;**614**:63-70

[18] Zhang C, Liu S, Qian G, Jian Z, Feng X. Effect of Sb content on properties of Sn–Bi solders. Transactions of Nonferrous Metals Society of China. 2014;**24**:184-191

[19] Ma Y, Li X, Zhou W, Yang L, Wu P. Reinforcement of graphene nanosheets on the microstructure and properties of Sn58Bi lead-free solder. Materials and Design. 2017;**113**:264-272

[20] Yan L, Lee C, Yu D, Yu A, Choi W, Lau J, Yoon S. A hermetic seal using composite thin-film In/Sn solder as an intermediate layer and its interdiffusion reaction with Cu. Journal of Electronic Materials. 2009;**38**:200-207

[21] Chen S, Lee W, Hsu C, Yang C, Hsu H, Wu H. Sn–In–Ag phase equilibria and Sn–In–(Ag)/Ag interfacial reactions. Materials Chemistry and Physics. 2011;**128**:357-364

[22] Li J, Mannan S, Clode M, Whalley D, Hutt D. Interfacial reactions between molten Sn–Bi–X solders and cu substrates for liquid solder interconnects. Acta Materialia. 2006;**54**:2907-2922

[23] Islam M, Chan Y, Rizvi M, Jillek W. Investigations of interfacial reactions of Sn–Zn based and Sn–Ag–Cu lead-free solder alloys as replacement for Sn–Pb solder. Journal of Alloys and Compounds. 2005;**400**:136-144

[24] Tian F, Liu Z, Shang P, Guo J. Phase identification on the intermetallic compound formed between eutectic Sn–In solder and single crystalline Cu substrate. Journal of Alloys and Compounds. 2014;**591**:351-355

[25] Li MY, Yang HF, Zhang ZH, Gu JH, Yang SH. Fast formation and growth of high-density Sn whiskers in Mg/Sn-based solder/Mg joints by ultrasonic-assisted soldering: Phenomena, mechanism and prevention. Scientific Reports. 2016;**6**:27522

[26] Zhao N, Zhong Y, Huang ML, Ma HT, Dong W. Growth kinetics of Cu_6Sn_5 intermetallic compound at liquid-solid interfaces in Cu/Sn/Cu interconnects under temperature gradient. Scientific Reports. 2015;**5**:13491

[27] Huang M, Yang F. Size effect model on kinetics of interfacial reaction between Sn–xAg–yCu solders and Cu substrate. Scientific Reports. 2014;**4**:7117

[28] Kumar KM, Kripesh V, Tay AA. Single-wall carbon nanotube (SWCNT) functionalized Sn–Ag–Cu lead-free composite solders. Journal of Alloys and Compounds. 2008;**450**:229-237

[29] Kanlayasiri K, Ariga T. Physical properties of Sn–58Bi–xNi lead-free solder and its interfacial reaction with copper substrate. Materials and Design. 2015;**86**:371-378

[30] Mu W, Zhou W, Li B, Wu P. Janus-faced Cu-core periphery formation and Bi phase redistribution under current stressing in Cu-cored Sn–58Bi solder joints. Journal of Alloys and Compounds. 2014;**584**:483-486

[31] Altıntas Y, Kaygısız Y, Öztürk E, Aksöz S, Keşlioğlu K, Maraşlı N. The measurements of electrical and thermal conductivity variations with temperature and phonon component of the thermal conductivity in Sn–Cd–Sb, Sn–In–Cu, Sn–Ag–Bi and Sn–Bi–Zn alloys. International Journal of Thermal Sciences. 2016;**100**:1-9

[32] Zhang S, Chen Q. Fabrication of MWCNT incorporated Sn–Bi composite. Composites Part B: Engineering. 2014;**58**:275-278

[33] Ping W. Effects of Zn addition on mechanical properties of eutectic Sn–58Bi solder during liquid-state aging. Transactions of Nonferrous Metals Society of China. 2015;**25**:1225-1233

[34] Aksöz S, Maraşlı N, Keşlioğlu K, Yıldız F. Variations of thermal conductivity with temperature and composition of Zn in the Bi–[x] at.% Zn–2 at.% Al alloys. Thermochimica Acta. 2012;**547**:1-5

[35] Sun H, Chan Y, Wu F. Effect of CNTs and Ni coated CNTs on the mechanical performance of Sn57.6Bi0.4Ag BGA solder joints. Materials Science and Engineering A. 2016;**656**:249-255

[36] Li X, Ma Y, Zhou W, Wu P. Effects of nanoscale Cu_6Sn_5 particles addition on microstructure and properties of Sn–Bi solder alloys. Materials Science and Engineering A. 2017;**684**:328-334

[37] Mokhtari O, Nishikawa H. The shear strength of transient liquid phase bonded Sn–Bi solder joint with added Cu particles. Advanced Powder Technology. 2016;**27**:1000-1005

[38] Li Y, Chan Y. Effect of silver (Ag) nanoparticle size on the microstructure and mechanical properties of Sn–58Bi–Ag composite solders. Journal of Alloys and Compounds. 2015;**645**:566-576

[39] Hu T, Li Y, Chan Y, Wu F. Effect of nano Al_2O_3 particles doping on electromigration and mechanical properties of Sn–58Bi solder joints. Microelectronics Reliability. 2015;**55**:1226-1233

[40] Wang F, Zhou L, Wang X, He P. Microstructural evolution and joint strength of Sn–58Bi/Cu joints through minor Zn alloying substrate during isothermal aging. Journal of Alloys and Compounds. 2016;**688**:639-648

[41] Li Y, Luo K, Lim AB, Chen Z, Wu F, Chan Y. Improving the mechanical performance of Sn57.6Bi0.4Ag solder joints on Au/Ni/Cu pads during aging and electromigration through the addition of tungsten (W) nanoparticle reinforcement. Materials Science and Engineering A. 2016;**669**:291-303

[42] Shalaby RM. Effect of silver and indium addition on mechanical properties and indentation creep behavior of rapidly solidified Bi–Sn based lead-free solder alloys. Materials Science and Engineering A. 2013;**560**:86-95

[43] Yang L, Zhou W, Liang Y, Cui W, Wu P. Improved microstructure and mechanical properties for Sn–58Bi solder alloy by addition of Ni-coated carbon nanotubes. Materials Science and Engineering A. 2015;**642**:7-15

[44] Peng H, LÜ X, Lin T, LI H, Jing A, Xin M, FENG J, Zhang Y, Qi L, QIAN Y. Improvement of mechanical properties of Sn–58Bi alloy with multi-walled carbon nanotubes. Transactions of Nonferrous Metals Society of China. 2012;**22**:s692-s696

[45] Ma D, Wu P. Effects of coupled stressing and solid-state aging on the mechanical properties of graphene nanosheets reinforced Sn–58Bi–0.7Zn solder joint. Materials Science and Engineering A. 2016;**651**:499-506

[46] Peng Y, Deng K. Study on the mechanical properties of the novel Sn–Bi/Graphene nanocomposite by finite element simulation. Journal of Alloys and Compounds. 2015;**625**:44-51

[47] Wojewoda-Budka J, Zięba P. Formation and growth of intermetallic phases in diffusion soldered Cu/In–Bi/Cu interconnections. Journal of Alloys and Compounds. 2009;**476**:164-171

Chapter 3

Brazing Behaviour of Ag-Cu Filler Materials

Ionelia Voiculescu

Additional information is available at the end of the chapter

http://dx.doi.org/10.5772/intechopen.69862

Abstract

Ecological brazing rods (cadmium free) represent a technical solution particularly useful for joining metallic or non-metallic components, providing the required mechanical and functional characteristics obtained in reproducible manufacturing conditions and at reasonable costs. The new class of coated rods for brazing must provide high deposition efficiency, chemical compatibility in relation to a number of metals and alloys currently used in industry and high corrosion resistance in different media. Such diverse characteristics can be obtained by achieving a special coating that contains a mixture of materials having a role of chemical activation and catalyst effects, as well as contributing to increasing the adherence to unmolten interfaces. This chapter presents some results obtained by using experimental brazing filler materials for different types of materials and applications. There are briefly highlighted some aspects on the diffusion effects of chemical elements in the soldering interface, the bonding of ceramics and some issues related to the effects of chemical elements from the brazing material.

Keywords: brazing, filler material, coated rod, diffusion, ceramics

1. Introduction

Currently, there is a range of alloys available on the specialized market, with different chemical compositions, used as filler materials for brazing. Of these alloys, a Cu-Ag class is used to obtain similar or dissimilar joints, operating in moderate corrosive environments at low or high temperatures. The Cu-Ag class brazing alloys yield specific characteristics such as high fluidity and ability to spread fast in very narrow interstices, high ductility and mechanical resistance, chemical stability in different environments and temperatures, good adhesion and excellent wetting capacity, suitable for a wide variety of metal or metal-ceramic materials [1, 2]. To determine the optimal operating temperature (melting point and eutectic temperature), the mutual solubility of chemical elements and the types of main phases formed (such as solid solutions, eutectic phases, compounds, etc.), see the phase diagrams of the alloying system (binary or ternary diagrams) in **Figure 1**.

In **Figure 1(a)**, the eutectic point is at about 780°C with the specific concentrations of 72 wt% Cu and 28 wt% Ag, whereas in **Figure 1(b)**, other values for temperature and chemical composition are shown. The differences between the values specified by different charts are determined by the different purity values of the elements and by the accuracy of determining the transformation temperatures. If other chemical elements are added in the metallic matrix of the filler material, in order to obtain some special characteristics (lower temperature of the eutectic point, narrower or wider solidification domain), it becomes difficult to determine the exact value of the eutectic point position. The temperature range for Cu-Ag class brazing alloys is between 500 and 1000°C [2]. These alloys are used for brazing parts made of carbon steel, galvanized steel or stainless steel, some copper alloys (Cu, Cu-Sn and Cu-Zn) or others. Since December 2011, the introduction of Cd in these brazing alloys has been banned, in accordance with the EU regulation 494/2011.

Besides Cu (22–40 wt%) and Ag (25–56 wt%), brazing alloys may contain elements such as Sn (2–2.5 wt%) and Zn (17–35 wt%). The melting domain is 620–790°C, depending on the chemical composition [2].

The advanced brazing filler materials contain some chemical elements added in the coating or in the brazing rods, which give them a great capacity for the wetting of metallic surfaces coated with oxide layers. At the lowest temperature value of the melting domain (650–700°C), the ceramic mixture from the coating starts to melt, resulting in deoxidation of the surface for the parent material. At higher temperatures, between 700 and 800°C, the second ceramic coating, containing a mixture of silver particles, starts to melt, completing the final chemical composition of the filler material. In this way, the filler metal can spread rapidly and it can completely fill the gap between the joined components [3–6]. The most used filler material brands are 244 Ag, 134 Ag and 145 Ag, which are designed to perform connections for drinkable water pipes

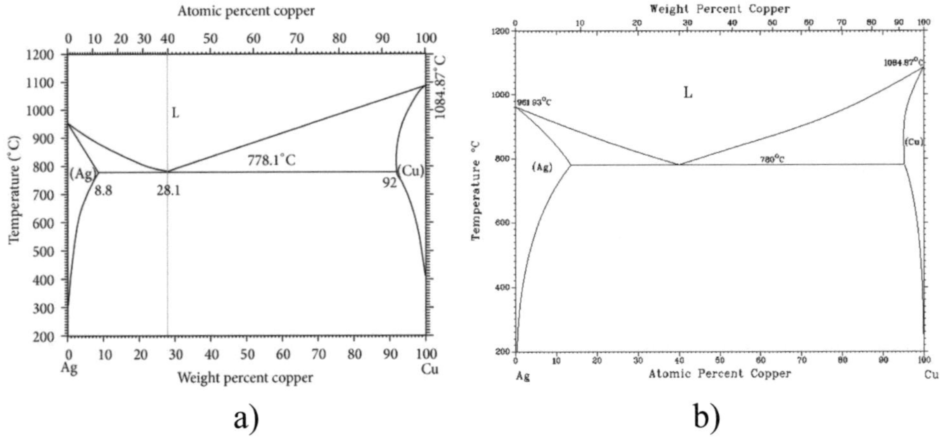

Figure 1. Phase diagram for the Cu-Ag alloying system. Cu: 71.9%wt, Ag: 28.1%wt [1, 2]; a) Melting Temperature: 778.1°C b) Melting Temperature: 780°C

with brazing temperatures of approximately 650–830°C, bundles or tubular exhaust systems. The operating conditions involve corrosive effects and oxidation at high temperatures (300°C), tensions and contractions in the joint. For these applications, the increase in the tin content can improve the wettability and mechanical properties of the joints. However, when the tin content exceeds 5% Sn, the shear strength of the joints decreases [7]. Lower temperatures desired for the diffusion brazing of stainless steels are not determined only by the melting point of the filler materials. In this case, the end product of the reaction between the filler material and parts is a solid solution, and residual interfacial tensions do not occur if the inter-metallic phases are not present. For that reason, the solubility limit of the minor metal constituents in the primary phase is crucially important [8].

Generally speaking, the stainless steel surface is wetted with difficulty by some brazing alloys, because the superficial oxide (chromium oxide) protects it against corrosion. Therefore, the chemical composition of the brazing flux is designed to ensure a rapid dissolution of oxides and to ease the wetting of the molten filler material on the base material. Due to the fact that additional surfaces of the brazing zone are affected by the dissolution effects favoured by the brazing fluxes, it is important to rapidly clean the surrounding surfaces after brazing [9]. The brazing flux preferred for this type of filler material is FB 10 [2, 3]. To ensure a good mechanical resistance of the joint, the brazing alloy must be sufficiently fluid, in order to penetrate the gaps between the components, on large distances. Especially in the case of dissimilar joints, the melting temperature of the brazing alloy that contains silver and copper must be well correlated with the type of base material. In the case of austenitic stainless steels, the holding in the 600–800°C temperature range is critical to avoid the inter-crystalline corrosion. In the case of very thin components, the effects of high temperatures are more critical [6].

This chapter presents some results obtained by using experimental brazing filler materials (ecological brazing rods, cadmium free, obtained under reproducible manufacturing conditions and at reasonable costs, metallic glass filler material and metallic foils) for different types of materials and applications.

The new class of coated rods for brazing can provide high deposition efficiency, chemical compatibility in relation to a number of metal and alloys currently used in industry and high corrosion resistance in different media. Such diverse characteristics were obtained by achieving a special coating that contains a mixture of materials having chemical activation role and catalyst effects, as well as contributing to increasing the adherence to unmolten interfaces. There are highlighted some aspects on the diffusion effects of chemical elements in the soldering interface of stainless steel 304 used for high temperature applications, dissimilar bonding of ceramics (tungsten carbide with steel), alumina bonding using metallic foil and some issues related to the effects of chemical elements from the brazing material. Some other results refer to the possibilities of joining ceramic materials using the various types of filler metal alloys.

2. Brazed joints

Several types of connections were made between different types of materials, such as 304 stainless steel, galvanized steel, carbon steel, tungsten carbide, ceramics pills (alumina) and copper.

As a general rule, before brazing, all the surfaces were cleaned with propane and then dried with hot air. After brazing, the joints were cleaned using hot water (90°C) and sampled for analysis. The samples were cut using a precision cutting machine that includes the Smartcut system to automatically adjust the feed rate and to prevent sample deformation.

2.1. Stainless steel brazing

To analyse the joint between very thin parts made of stainless steel, the spreading mode in the brazing gap and the diffusion of chemical elements in transition zones from the heat-affected zone (HAZ), two types of joints were analysed: corner joint (**Figure 2a**) and overlap joint (**Figure 2b**). The parts made of 304 stainless steel type have 0.5 mm thick, with a relative gap of 0.4 mm between them. Brazed joints were obtained using the oxy-flame process, with a

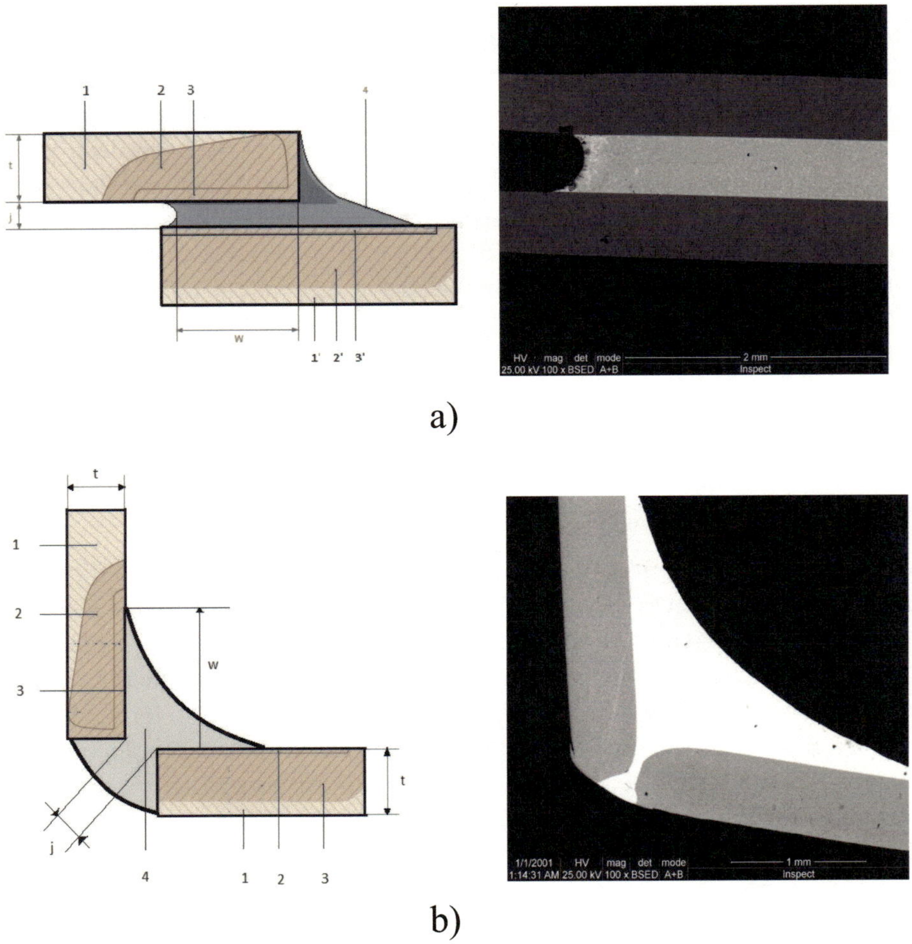

Figure 2. Schematic diagram and images for brazed joints: 1 and 1′—parent material unaffected by the brazing process; 2 and 2′—heat-affected zone; 3 and 3′—diffusion zone/transition zone; 4—braze metal; *j*—actual joint width; *w*—overlap/lap length; *t*—parent material thickness. (a) overlap brazed joint (100×) and (b) corner brazed joint (100×).

slightly oxidizing flame and VIAg22SnSiPR filler material (max. 22 wt% Ag; max. 43 wt% Cu, max. 35 wt% Zn, 0.5–1 wt% Sn and 0–0.2 wt% Si) [3].

The diffusion length for the chemical elements in the brazing zone was measured both in the base material and in the heat-affected zone using the EDX - Energy-dispersive X-ray spectroscopy analysis. The total duration of the brazing process was of 30 s, including the heating of components and the actual brazing, in order to avoid the deformation of the thin components and the corrosion effects. Cooling was carried out in cold water immediately after brazing. The samples were washed with hot water and dried in hot air. The representative areas from each brazed sample were cut using a cutting machine with a cooling agent to avoid the alteration of the microstructure [3]. The 'overlap' joint (**Figure 2a**) has the average size width (*j*) of 412 μm. The cross-section aspect of the brazing zone is correct, without molten material excess and with well-formed concavity, meeting the conditions imposed for this type of joint. The condition for the required maximum joint resistance is to have a minimum overlapping length of 5 *t*, where '*t*' is the thickness of the components. In the analysed case, the actual overlapping length (*w*) was of 4 mm, higher than the required minimum value of 2.5 mm. In the case of the 'corner' joint (**Figure 2b**), the geometrical dimensions of the brazed zone were actual joint width (*j*) of 42 μm in the root zone and maximum overlapping length (*w*) of 3.14 m. Individual measurements (EDX analysis) of the chemical composition were taken in 5 points at different distances from the brazing surface (**Figure 3a**).

The EDX measurements were afterwards performed in the molten filler material volume, by individual measurements taken in 5 points of the cross-section of the 'overlap' joint (**Figure 3b**). The results of each chemical composition measurement are shown in **Tables 1** and **2**.

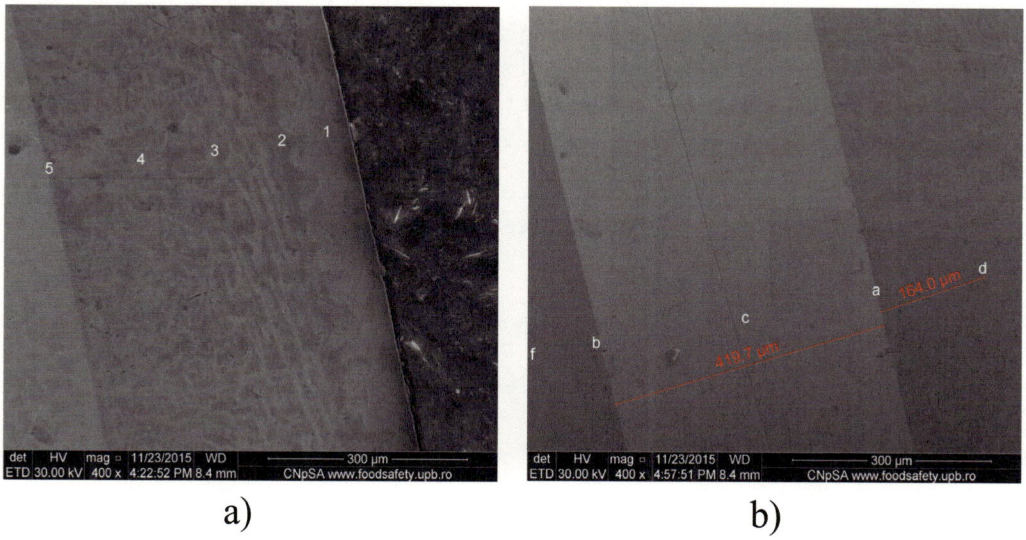

Figure 3. Location of the measurement points (overlap joint): (a) parent material; (b) filler material [3]; (a) 400× and (b) 400×.

Measurement zone	Chemical composition (wt%)									
	Si	Cr	Mn	Fe	Ni	Al	Ag	Zn	Sn	Cu
1. Base metal	0.23	18.82	2.5	70.94	7.5	–	–	–	–	–
2. HAZ	0.2	18.08	2.27	69.34	6.92	3.18	–	–	–	–
3. HAZ	0.19	18.16	2.46	69.53	7.59	2.06	–	–	–	–
4. HAZ	0.23	18.48	2.49	69.83	7.14	1.83	–	–	–	–
5. Near fusion line (at 20 μm)	0.3	13.45	1.56	49.43	4.63	–	3.95	10.97	0.43	15.28

Table 1. Chemical composition measured in 5 points in the cross-section of the parent material (**Figure 3a**) [3].

As shown by the chemical composition values in **Table 1**, diffusion effects are detected only at very small distances from the molten metal interface, at length of approximately 20 μm. There can be seen the significant diffusion effects of elements such as silver, zinc and copper, from the filler metal to the parent material (**Table 1**, point 5, **Figure 3a**) and the diffusion effects of chromium, iron and nickel starting from the parent material to the molten material.

To estimate the diffusion effects from the filler material to the base material, measurements were carried out outside the transition zones (**Figure 3b** and **Table 2**), at lengths of around 160 (point 'd') and 60 μm (point 'f'), respectively. The measurement points 'a' and 'b' were located in the transition zone, between the parent materials (overheated) and the molten filler material, approximately 20 μm away from the fusion line. Point 'c' corresponds to the chemical composition in the centre of molten filler material (around 200 μm from the interface with the parent material).

The most intensive diffusion effects were obtained in the case of copper (over 25%), zinc (over 21%) and silver (12%). To evaluate the global effects of diffusion (corner joint), the analysis of the chemical composition was made using scanning electron microscopy in 105 different points, located in the base material, root and seam zone from **Figure 4**. The measurements were carried out in straight line, from left to right, between points 1 and 105, at a distance between successive measurements of 10 μm.

The measurement distance in the first and second base material was of maximum 240 μm, on both sides of the fusion line. As it can be seen, the transition from the filler material to the

Measurement zone	Chemical composition (wt%)								
	Si	Cr	Mn	Fe	Ni	Ag	Zn	Sn	Cu
a. Base metal	0.28	15.12	1.58	56.99	4.92	3.19	6.99	0.21	10.7
b. Base metal	0.09	14.11	1.54	56.75	5.71	1.96	7.57	0.13	12.15
c. Central zone of filler metal	–	–	–	–	–	21.36	34.5	1.16	42.98
d. HAZ (164 μm)	0.23	18.43	2.74	70.33	8.1	–	–	–	–
f. HAZ (60 μm)	0.22	18.45	2.33	69.24	7.28	–	–	–	–

Table 2. Chemical composition measured in 5 points through filler material and HAZ (**Figure 3b**) [3].

http://dx.doi.org/10.5772/intechopen.69862

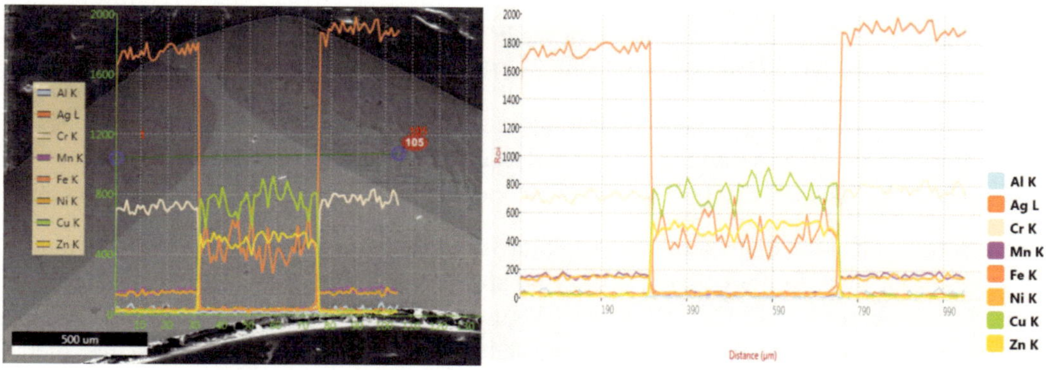

Figure 4. Evolution of the chemical composition in 105 different points for the corner joint [3].

base material is in a very narrow zone, the diffusion effects being visible only at 20 μm from the fusion line.

The metallographic analysis of the brazed samples targeted the visualisation of the molten alloy distribution in the joint's gap, the correctness of the molten filler material spreading and the absence of imperfections. The microstructural analysis was conducted using a scanning electron microscopy provided with electron gun with field emission—EGF with a resolution of 1.2 nm and an X-ray energy dispersive spectrometer (EDS) with a resolution of 133 eV at MnK.

In the case of the 'overlap' joint, a good adhesion between the brazing filler metal and the stainless steel surface was obtained (**Figure 5a**), despite a few pores located near the interface with the

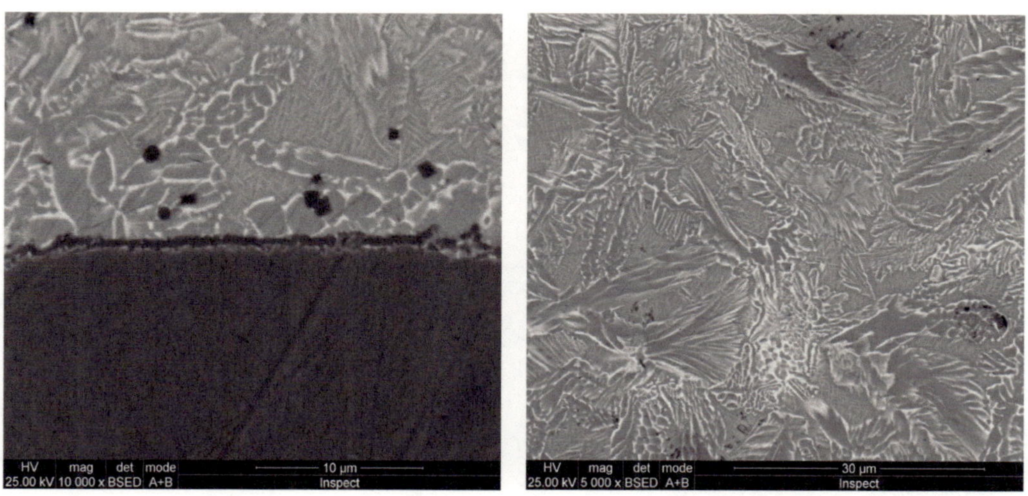

a) b)

Figure 5. Microstructure appearance in the cross-section of the overlapped brazed joint: (a) interface between filler material and parent material; (b) filler material. (a) 10,000× and (b) 5000×.

parent material. The phases present in the molten metal have needle-like shape (**Figure 5b**), due to the cooling conditions. The expansion of the heat-affected zone (HAZ) was quite restricted (about 315 μm), given the very small thickness of the stainless steel components (0.5 mm).

In the case of the 'corner' joint, at higher magnification degrees (**Figure 6a**), there can also be seen a good adhesion between the molten material and the stainless steel surface. The correct geometry of the root and the absence of the secondary fragile phases are visible (**Figure 6b**).

Some pores with small diameters (about 1–2 μm) are located near the interface, with random spreading, but they do not endanger the bond strength (0.5 mm diameters are admitted for imperfections according to EN 12799-2002) (**Figure 6b**). During tensile tests, the overlapping samples were broken in the base material, thus highlighting a good behaviour of the filler metal.

The experimental alloy used in the study has a good spreading behaviour and does not produce any cracking tendency. The only types of imperfections observed in the brazed zone were unmolten flux inclusions and pores (gas) resulted from the coating of the filler material.

In the brazed joints made by stainless steel components, at the interface between the base material and the molten filler material, a diffusion phenomenon can occur due to the migration of some chemical elements through the interface and even the holding time at high temperatures is very short [3].

2.2. Dissimilar brazed joint

The cooling of a dissimilar joint after brazing (i.e. steel brazed with tungsten carbide) is very important due to the high expansion coefficient of the steel that tends to contract faster than

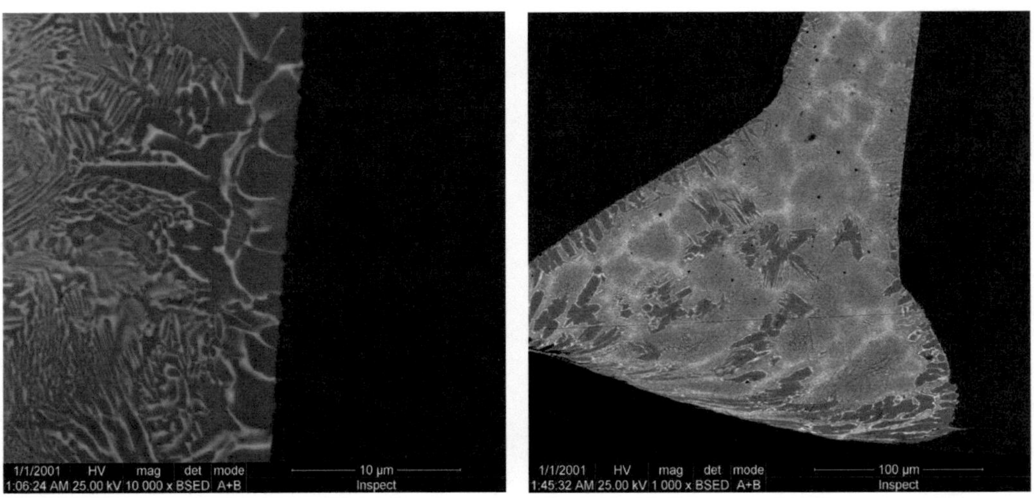

a) b)

Figure 6. Microstructure aspects in the cross-section of the brazed corner joint: (a) interface between the filler material and the parent material; (b) brazed root zone. (a) 10,000× and (b) 1000×.

the part of tungsten carbide to which it is brazed. This mismatch develops shear stresses within the interfaces of the joint.

If a high-strength brazing filler material is used, these stresses will be directly transmitted to the brittle tungsten carbide and could cause its cracking (**Figure 7**) [7, 10]. For this reason, the filler metal must have a low yield point to be able to deform and to allow the dissipation of the shear stresses. If the dimension of the tungsten carbide does not exceed 10 mm, the best solution is to use a filler metal like EN 1044 AG102, 103 and 303, which is sufficiently ductile and provides a thin layer of brazing filler metal. Another advantage of these filler metals is their low-brazing temperatures, helping to minimise the amount of differential expansion developed between the tungsten carbide and steel [10].

To ensure good conditions for penetration by capillarity of the brazing alloy, a proper preparation of the surface of components is required. The brazing of different (i.e. low alloy steel and tungsten carbide) filler materials must provide a good wetting for both types of surfaces, at the same time, at similar operating temperatures.

Any substance in the brazing gap or in the filler material can cause a change in the wetting behaviour, preventing fluid flow [10, 11]. If the material surface is too smooth, there is obtained an inhibition of the metal flow and poor adhesion (tungsten carbide has low roughness, as it is obtained by pressing and sintering). A rough surface can increase wettability, but in the case of higher values, it can affect the spreading mode of the molten filler material and can favour mechanical stress on the micro-surface irregularities [12].

Another problem that can occur during heating is the oxidation of the tungsten carbide component. The oxidation can occur on both surfaces causing superficial damage and the emergence of micro-cracks. The brazing surface must be completely protected against oxidation by the coating provided by paste or flux. When brazing tungsten carbide with steel component, the heat

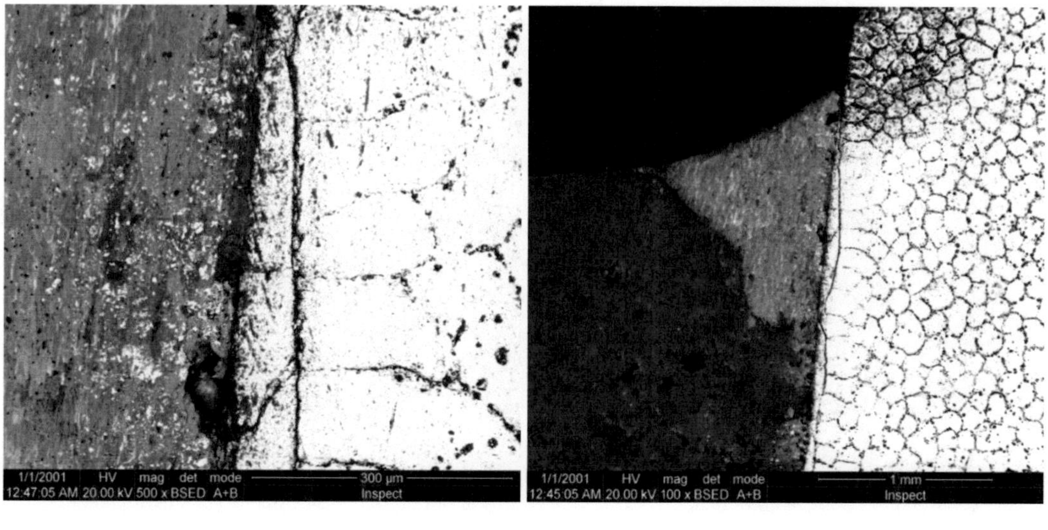

Figure 7. Crack formation in tungsten carbide brazed with steel components (500×, 100×) [7].

flow is more effective on the sides and less at the bottom of the joint, as the complete penetration of the brazing alloy in these zones is prevented. Therefore, to balance the heat flow, a few pieces of solder wire can be placed under the tungsten carbide to contribute to better warming and heat dispersion into the gap, by wetting the entire surface of the parts, from the bottom up (**Figure 8**).

Currently, there are several types of fillers for brazing tungsten carbide with steel, classified into two main groups.

The first group contains copper and copper alloys, including Cu-Zn (special brazing alloys) and Cu-Zn-Ni alloys. The first types of brazing alloys for tungsten carbide were those containing 100% Cu or 60% Cu + 40% Zn. Subsequently, other copper alloys have been developed by adding chemical elements such as Cd, Ag and Si. Copper is used not only because it is very malleable and can be processed as a sheet or a rod in different dimensions, close to those required (under 0.08 mm), but also because it allows the accommodation and dissipation of internal stresses during cooling. Note that 0.4-mm thick Cu-based sandwich sheets for brazing contain a middle sheet of 0.2 mm, covered by two 0.1-mm thick sheets.

The second group contains alloys with low melting point, containing silver in different proportions. The first brazing alloys based on silver have been developed since 1944 and they are mainly used for brazing tungsten carbide. Subsequently, 3% Ni or Cd was added in these alloys for low temperature applications. Adding Ni led to increase the carbide wetting capacity, as well as increase the capacity to accommodate tensions in very thin interstices of the joint. Nowadays, nickel is a common element added in filler materials for brazing WC.

Of the two main groups, Ag alloys are most commonly used because they have melting points below 750°C and exhibit an excellent brazing behaviour. Such brazing alloys are suitable for both induction and flame gas (**Figures 9** and **10**), applicable both for WC and steel, since their

a) b) c)

Figure 8. Assemblies prepared for brazing: (a) as-cast steel body and tungsten carbide tip; (b) solder wire placed in the brazing gap; (c) brazed assemblies [3, 7].

a) b) c)

Figure 9. Procedures for brazing dissimilar joints between steel and tungsten carbide: (a) oxy-flame brazing; (b) induction brazing; (c) brazed assembly [7].

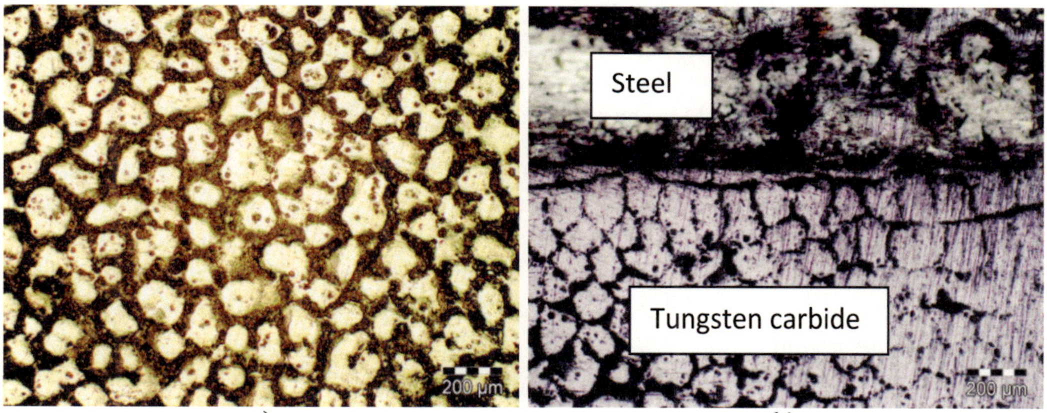

a) b)

Figure 10. Optical microscopy images of the joining zone obtained by brazing between tungsten carbide and steel: (a) tungsten carbide microstructure and (b) joint between tungsten carbide and steel [7]. (a) 200× and (b) 200×.

melting point is below the Curie point of steel. This characteristic is very important during brazing by induction, due to the fact that on reaching the Curie point, the steel becomes paramagnetic and changes the effectiveness of the inductive heating effect (it requires a larger amount of energy for the soldering process). For manual, the oxy-flame brazing process has been used as an experimental filler material Ag15CuP6 (cadmium free) (max. 15 wt% Ag, max. 82 wt% Cu, max. 3 wt% P) having 2 mm diameter and 450 mm length [7]. The stages of brazing process (**Figure 9a**) were as follows:

- positioning of the tungsten carbide tip into the brazing gap;
- protecting the free surface of the ceramic body using refractory slurry (dip coating, leak and drying in furnace at 150°C for 1 h);
- application of flux into the brazing gap;
- heating the assembly with flame, by moving the flame around the steel body until reaching a prescribed preheating temperature, without exceeding the melting temperature of the brazing alloy;
- continuing the heating by positioning the flame on the ceramic tip and adding filler material (**Figure 10**) until a slow refluxing in the brazing gap can be seen;
- slow cooling of the assembly in air;
- removal of the flow excess using hot water (40°C) and alcohol, followed by drying in hot air.

FH10 flux is generally used for brazing and soldering, according to EN 1045. The operating temperature range of this flux is 550–800°C, coupled with filler materials that have a melting point below 800°C (typically below 750°C). The condition required is for the selected flux to be active at minimum 50°C over the liquidus temperature of the brazing alloy. The flux is recommended for homogenous or heterogeneous joints, for components made of carbon or stainless steel, copper, brass, ceramic materials but not aluminium. It is recommended especially for the brazing of aluminium bronzes, steels, tungsten or molybdenum carbide or whenever protection is needed for components during heating [7, 9, 11].

The preheating stage during induction brazing (**Figure 9b**) must be carried out gradually (40–60 s) until reaching the melting temperature for brazing, using the following parameter values: 60 s brazing time, 7–8 A current, 60–175 V voltage and 30 kHz frequency. Otherwise, unwanted effects may occur, as brazing alloy boiling. If the assembly is overheated, it may cause the boiling of the brazing alloy, which can generate its expulsions outside the bonding zone or pores formation.

The heating process can be monitored using a thermal infrared camera in order to establish the exact temperature achieved during brazing. After brazing, cross-section of joining area can be evaluated using optical and SEM microscopy (**Figure 11**).

Recent filler materials used for the brazing of tungsten carbide are those alloyed with Ag and Zn (Cd-free), whose disadvantage is the high potential of evaporation during soldering.

For this reason, in the most recent filler materials, Zn is replaced by Sn, Cu, Ni, P and Mn (**Figure 12**), sometimes accompanied by small amounts of indium or tin, to decrease the melting temperature [13–17].

2.3. Ceramic joining

According to Schwartz [18] and Klomp [19], the continuous diffusion joint between ceramics and metal is greatly influenced by the surface quality of the filler material because it involves

a) b)

Figure 11. Optical microstructure in cross-sections of brazed joint (a) and SEM microstructure of filler material (b). (a) 500× and (b) 800×.

the ability to get into very close contact with the entire free ceramic surface. If the ceramic surface is smooth, the time required for bond formation at the atomic level is determined by the deformation of the metal. Surface filler material irregularities can prevent the formation of a continuous interface, even by applying a pressure, as superficial energy can change during plastic deformation [13]. The ceramic surface quality depends on the processing method. Therefore, in the experiments, alumina pills were polished using sandpaper with different grits, until a smooth surface was obtained (a roughness of about 0.8 μm) [20, 21].

The researcher analysed the joint between two components of Al_2O_3 using a metallic interlayer placed between the contact surfaces of ceramic. There were used several types of filler materials, including Co-Nb-Si-B (0.1 mm), FeNiB (0.1 mm), Cu 99% (0.2 mm), Al 99.99%

Figure 12. Experimental filler material (B Ag 40CuSnP, Cd-free) and manufacturing factory [21].

(1 mm), $Fe_{73.5}Cu_1Nb_3Si_{15.5}$ metallic glass (0.01 mm), CoNbSiB (0.15 mm) and $Cu_{50}Ag_{40}Zn_{10}$ (0.8 mm). Brazing was performed using an electric furnace in a controlled atmosphere. The brazing parameter values used for the experiments were as follows: the maximum temperature of 1100°C, 60-min holding time at maximum temperature, 5°C/min heating and cooling velocity in an argon atmosphere.

After joining, the assembled parts were mechanically tested (shear test) and the microstructure of both fracture surfaces and cross-sections was analysed in order to evaluate the integrity of the joint, the diffusion of the chemical elements and the nature of the compounds formed into the interface between the ceramic and metallic materials [20].

The values of the chemical composition were measured in different points located in each zone of interest by X-ray microanalysis. A common feature of all ceramic samples combined using various metal foils is that there can be seen, in their interface zone, the appearance of a semi-vitreous material with complex chemical composition, which tends to expand to the metallic material (filler material). In zones where the semi-vitreous material is formed in sufficient quantity, the adhesion of the ceramic parts is good and this material is behaving like a binder. This new type of material is probably the result of chemical reactions that took place between the filler material, ceramics and the pickling solution because it contains chemical elements of all three. Adhesion is poor in zones where this semi-vitreous material is missing or is insufficient.

During experiments, it was found that, in the case of very soft metals (1000 μm-thick Al foil), poor adherence was obtained during soldering due to excessive deformation of the metal foil between the surfaces to be joined. When 0.2-mm thick foil was used, there were obtained better results (a breaking shear of about 85 N/mm^2). The use of 0.1-mm thick FeNiB alloy sheets yielded poor shear resistance (around 70.3 N/mm^2) while 0.01-mm thick $Fe_{73.5}Cu_1Nb_3Si_{15.5}$ metallic glass strips yielded a much higher value of shear resistance (267 N/mm^2) [21]. The analysis by electron microscopy and X-ray diffraction showed that an intermediate metallic layer was obtained at the interface, leading to the conclusion that the diffusion bonding mechanism occurred most likely during brazing.

The microstructural appearance seen in the case of the sample joined using the CoNbSiB metallic glass was as follows: the pores of the ceramic material were filled with semi-vitreous materials in the zones with complete adhesion, evidencing a transition zone with own chemical composition, similar to a mixture between ceramic and filler materials (containing elements such as Co, B, Si and Nb).

Fine cracks are visible in the metallic filler material, separating volumes of materials by a few hundred nanometres. The transition zone has compact appearance, free of pores or cracks (**Figure 13b**). The results of the chemical composition (**Table 3** and **Figure 14**) show a transition zone between the ceramic material (points 1 and 2) and the metallic material (points 5 and 6) like a new semi-vitreous material, containing elements that diffuse both from the ceramic (Al and O) and from the filler material (Co, Nb, Si and B). Ancillary elements coming from surface preparation solutions are also present in the vitreous materials (Na, Mg, K and Ca).

Part of the metallic material was spun off during the shear test, due to its weak adherence during brazing. Breaking was located mainly at the level of the semi-vitreous material, which

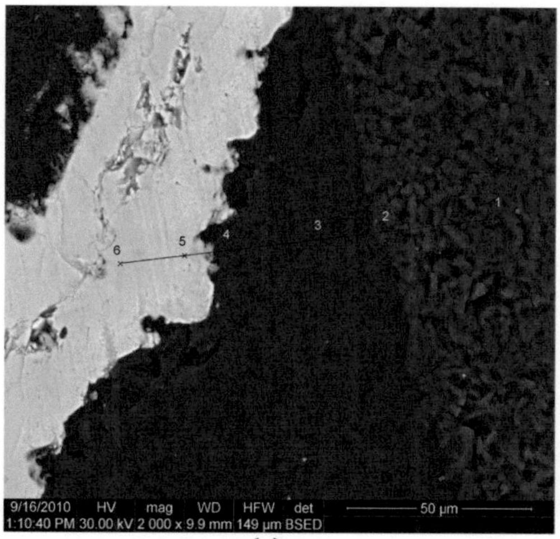

(a)

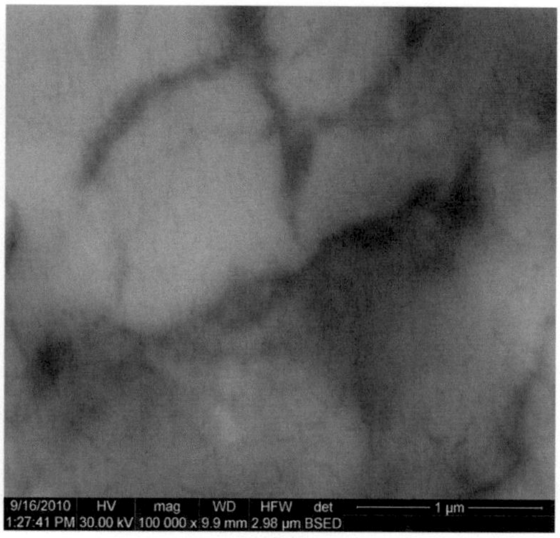

(b)

Figure 13. The melting temperature is situated in range of 778–780°C, depending on the chemical composition and accuracy of chemical composition method [1, 2]. (a) overall cross-section (ceramic, transition zone and metallic filler); (b) appearance of filler material (very fine cracks). (a) 2000× and (b) 100,000×.

shows a low deformation capacity compared with the metallic material and a lower mechanical strength compared to ceramics.

Bonding zones occur in the cross-section of the sample joined using FeNiB (**Figure 15a**), containing metallic material embedded in the semi-vitreous zones, forming a continuous belt and being deformed towards the interface with the ceramic materials. The diffusion of chemical elements forms some Fe-rich micro-crystals, 'surrounded' by thin films of semi-vitreous materials. In the case of the sample joined using the FeNiB metallic glass, the fracture surfaces obtained after shear tests are 40% covered by metallic materials. On the fracture surface, there were observed, in addition to the metallic material with a dendritic structure, some semi-vitreous zones in which very fine metallic polyhedral crystals were increased, having dimensions of 5–6 µm.

Point	Chemical composition (wt%)										
	Al	O	Si	Na	Mg	B	K	Ca	Fe	Nb	Co
1	60.77	31.29	5.02	2.04	0.23	–	–	0.65	–	–	–
2	25.46	39.12	8.51	9.46	0.62	5.7	0.40	3.17	–	6.89	–
3	24.82	39.7	8.92	9.12	0.61	6.57	0.25	3.31	0.24	5.46	0.73
4	25.22	40.39	9.85	8.87	0.65	5.45	0.39	3.27	0.13	3.69	2.10
5	2.53	–	0.92	2.61	–	3.3	–	–	0.70	5.18	84.76
6	2.73	–	1.16	3.12	–	3.0	–	–	0.37	15.43	74.20

Table 3. Chemical composition measured in 6 points through the cross-section (**Figure 13a**) [16].

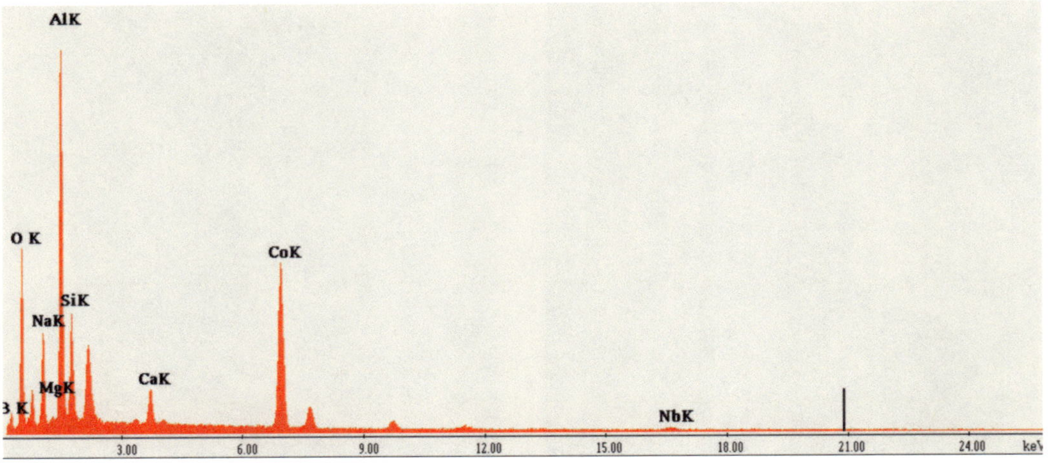

Figure 14. Energy spectrum for micro-zones from **Figure 13**.

Nanosized crystallites, which keep the nanostructured feature of the filler material, occur in zones covered by metallic material fragments (**Figure 15b**). Very fine cracks were observed both in the metallic and vitreous materials. Good adhesion to ceramic material is being provided by forming a semi-vitreous zone, with high iron content and Ni-free.

The main chemical elements present in this layer are as follows: Al, Fe, O, Na, Mg, Si and Ca (**Figure 16**).

The sample brazed with FeNbB alloy showed an asymmetry of the diffusion path in the ceramic layers, one of them containing large quantities of metal and polyhedral crystalline particles virtually several hundred microns deep (**Figure 17**).

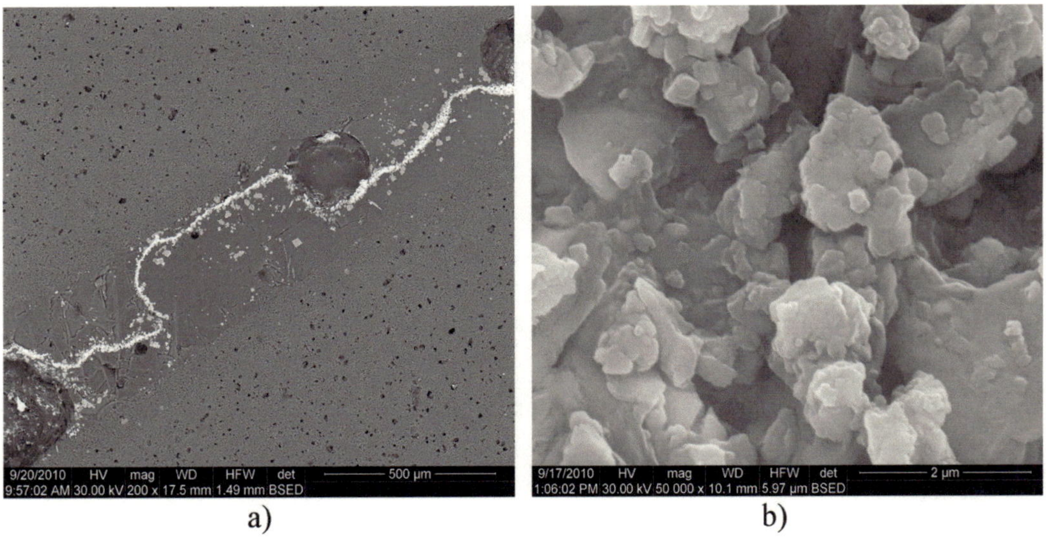

Figure 15. The melting temperature is situated in range of 778–780°C, depending on the chemical composition and accuracy of chemical composition method [1, 2]. (a) overall cross-section (ceramic, transition zone and metallic filler); (b) appearance of filler material in fracture surface (nanosized crystallites). (a) 200× and (b) 50,000×.

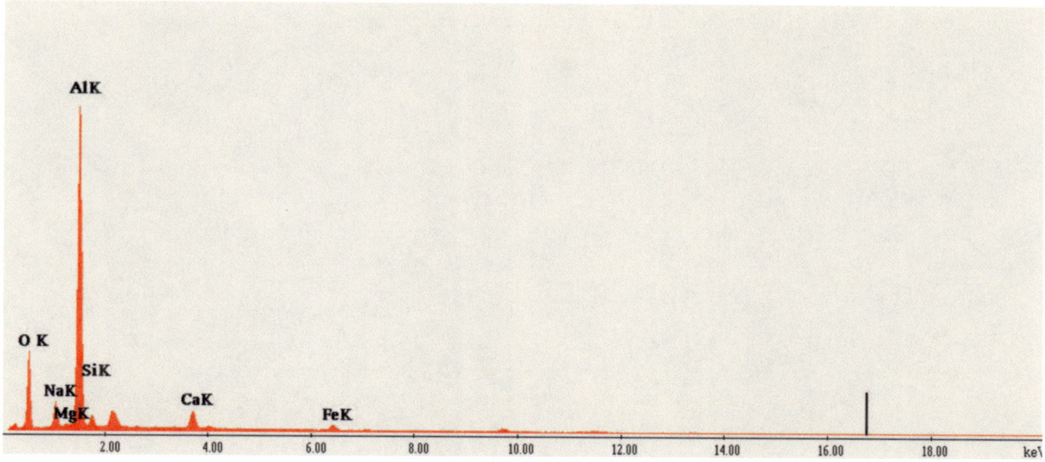

Figure 16. Energy spectrum for micro-zones from **Figure 17**.

In the case of the sample joined using the FeNbB metallic glass, the fracture surface resulted from the shearing test shows around 60% of the zone covered by metal (**Figure 18a**).

In the semi-vitreous mass (**Figure 18b**), there were observed metallic crystals containing mainly iron and silicon 'embedded' in a thin film of binder.

The microstructural appearances observed by scanning electron microscopy in samples joined using a copper foil as a filler material revealed melting-solidification effects in the metallic layer and a significant infiltration of liquid material through the porous ceramic materials. The entire surface of the ceramic layer is covered by a compact oxide layer. In the vicinity of

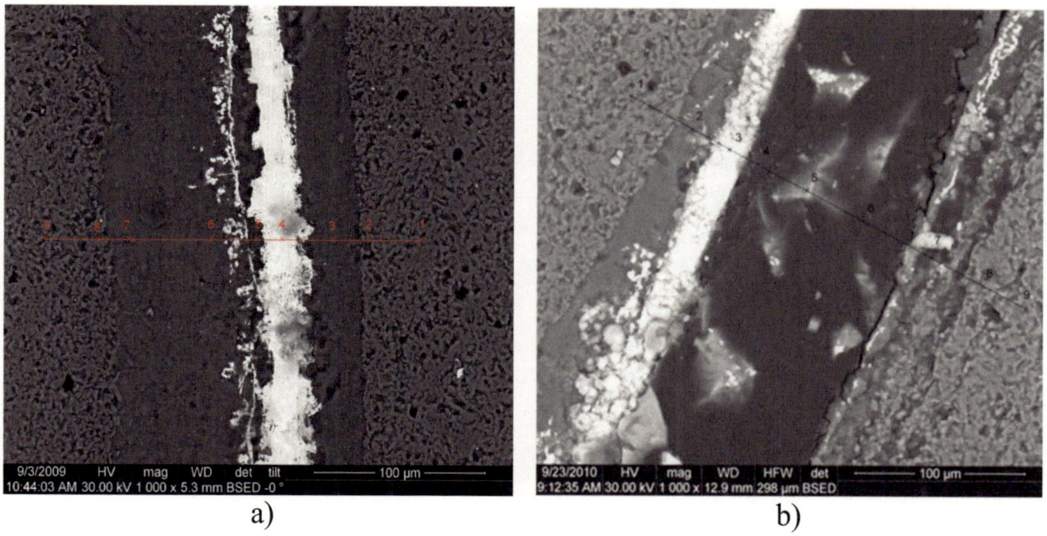

Figure 17. Cross-section of Al_2O_3 joint using FeNbB foil: (a) the joint appearance of the zone between two parts of aluminium oxide, containing a metallic layer in the central zone, bordered by two metal-ceramic zones; (b) vitreous zone containing metallic precipitates and having a porous structure. (a) 1000× and (b) 1000×.

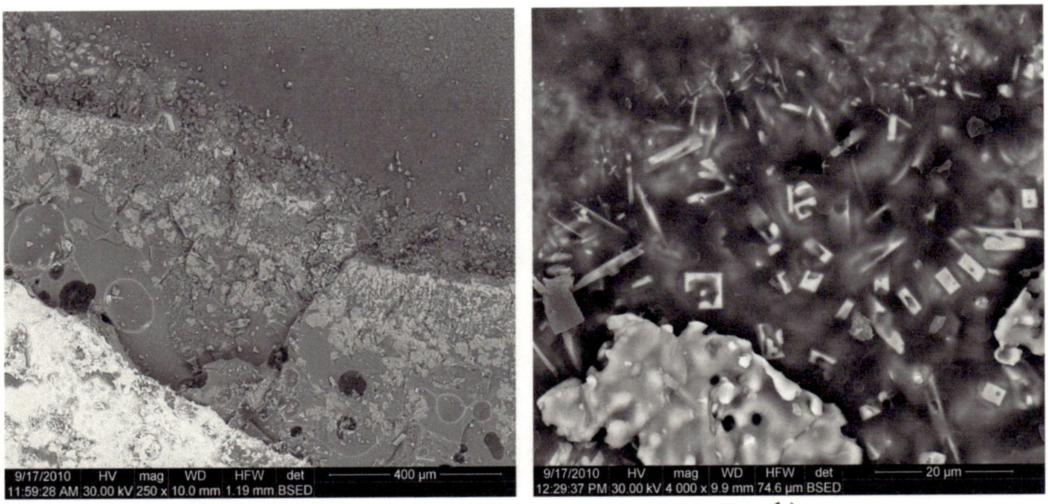

a) b)

Figure 18. Fracture surface resulted from the shearing test in the case of the sample joined using FeNbB metallic glass. (a) 250× and (b) 4000×.

the metallic foil, layers with multiple cracks are visible, rich in O, Al, Na, Ca and Si and containing lower amounts of Mg, C, K and Cl.

Their appearance suggests a mixture of oxides whose continuity is interrupted by Cu-rich and Si-rich compounds with polygonal morphology (probably separated from the liquid during solidification) (**Figures 19** and **20**).

Unlike the samples previously analysed, the use of copper as a filler material allows obtaining better adherence to ceramic, without discontinuity. Copper infiltrated the alumina pores

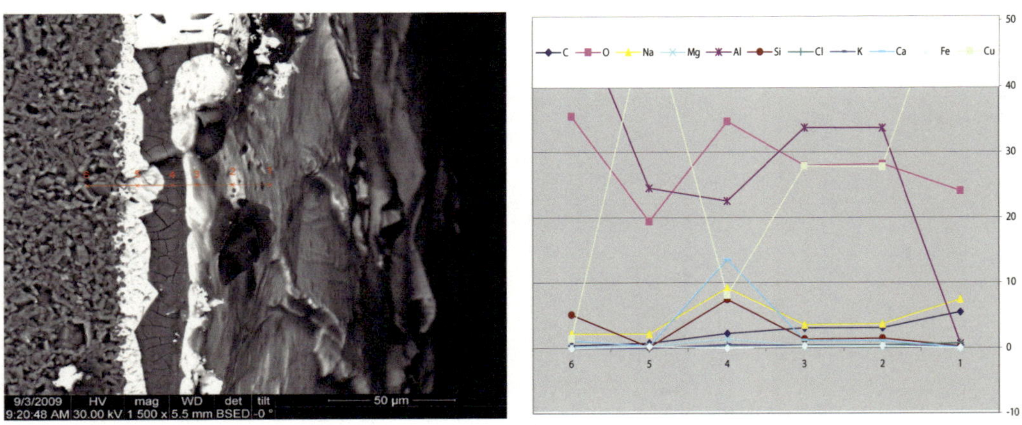

a) b)

Figure 19. Cross-section of Al_2O_3 joint using Cu foil (1000×) (a) and the evolution of chemical elements concentration in six different points from the cross-section of the Al_2O_3 joint using Cu foil (b) Distribution of chemical elements in the brazing interface, %wt: 24–34 wt% O; 0.63–53.96 wt% Al; 1.45–61.21 wt% Cu; 0.27–7.40 wt% Si. (a) 1500×; (b).

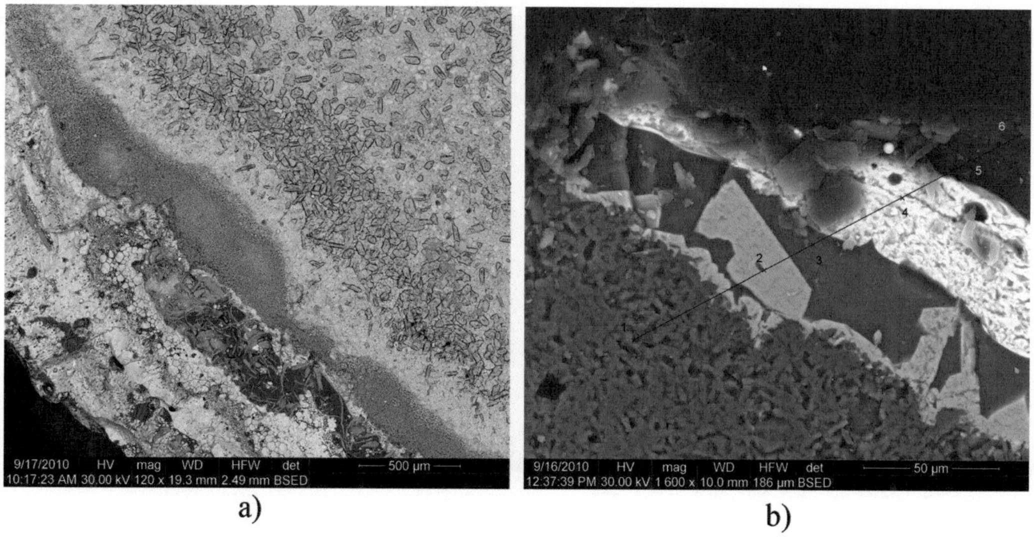

Figure 20. Fracture surface resulted from the shearing test in the case of the sample joined using copper foil. (a) 120× and (b) 1600×.

(about 2 mm deep), giving good mechanical resistance. The metallic material from the joining zone shows a dendritic structure, with very fine crystallites (tens of nanometres) (**Figure 20**).

Fracture mode resulted as peeling or pulling-out of the metallic layer from the ceramic material. An estimated 50–60% of the fracture zone is located in the metallic materials and some rare and adherent ceramic crystals that are visible. Some micro-cracks are also present on the fracture surface, generated by shear stress or by thermal contraction during the solidification of the filler materials. The metallic material was solidified in dendritic mode and contains polyhedral crystals of different sizes and geometries, influenced by the cooling rate. The microstructure also contains coarse crystals (several hundred microns in size) and very fine crystals (a few hundred nanometres in size), quickly solidified (**Figure 21**).

The shear strength of the joint is given primarily by the vitreous layer strength. The good adherence between the vitreous layer and the polygonal nanoparticles is provided by the good chemical compatibility and by the size of the particles. If the particle size is reduced, the resistance of the vitreous layer is increased. The metallic material is mechanically anchored on the porous ceramic layer and it adheres to the semi-vitreous material that contains some chemical elements diffused during brazing.

The influence of the reactive nature of the braze material on the joint integrity and strength was evaluated using shear tests. Cylindrical samples were used to conduct a shear test. A simple device and the LLOYD Instruments testing machine were used in the shear test. The normal force applied in the centre of the sample was balanced by the shear force, so the shear stress can be located at the level of the metallic material. The peak stress in the plot corresponds to the peak load at which shear failure occurred in the joint.

Shear failure occurred entirely in the braze material. As shown by the shear test (**Figure 22**), the best values for mechanical resistance are provided by a very thin layer of semi-vitreous material (ferrous-based metallic glass) [20].

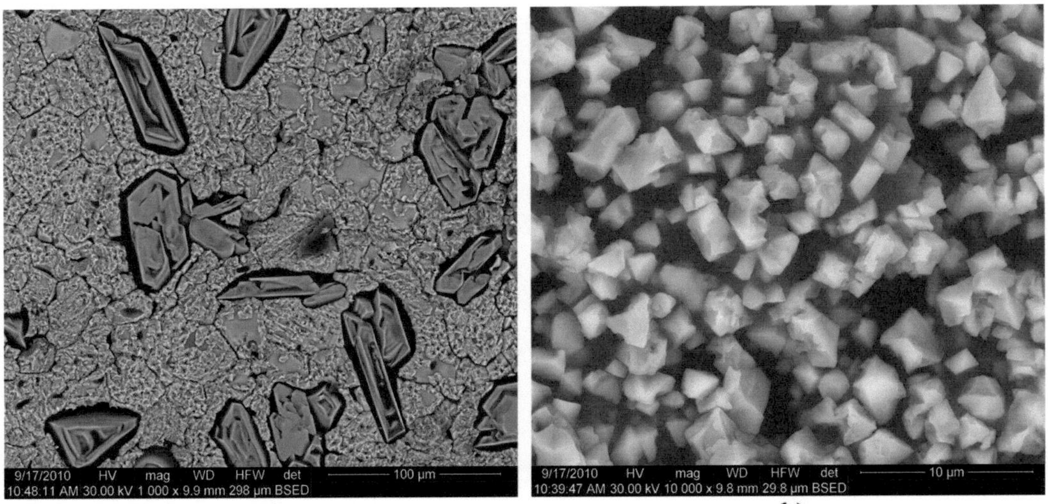

a) b)

Figure 21. Polyhedral crystals of different sizes and geometries in the sample joined using copper foil. (a) 1000× and (b) 10,000×.

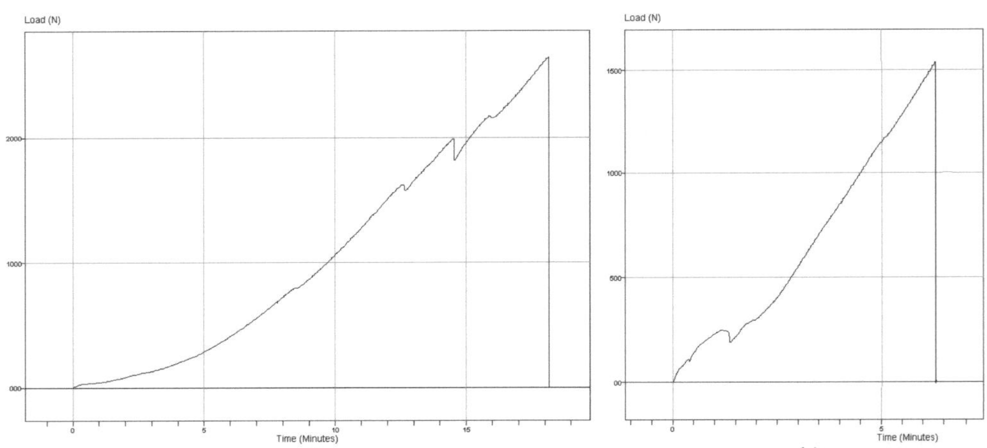

a) b)

Figure 22. Schematic shear test diagrams: (a) ferrous-based metallic glass. F_{max}=2670 N, testing speed v= 0.2 mm/min and (b) copper [14]. F_{max}=1534 N, testing speed v= 0.1 mm/min.

3. Conclusions

The brazing of parts made up of stainless steels using fillers containing Cu and Ag requires specific operating conditions in order to avoid contamination with chemical elements that can cause embrittlement or hot cracking. It is therefore necessary to apply brazing procedures with very short operating times, utilizing filler materials with narrow domain of melting and solidification.

Furthermore, the fluxes used for this application must ensure the pickling effect on the brazed surfaces at temperature 50°C above the melting temperature of the filler material.

The elements with greater participation in the filler material (copper, zinc and silver) migrate by diffusion towards the base material, the biggest rate of diffusion being for copper (25%). These diffusion effects are extending only in a small area, of 20 to 60 microns form de interface, and do not affect the mechanical resistance or corrosion resistance of the base material.

During the brazing of tungsten carbide with steel parts, there must be ensured a good penetration of the molten filler material into the brazing gap, given that the wettability of the two types of material is different and problems related to the bond strength can occur. Recent filler materials used for the brazing of tungsten carbide are those alloyed with Ag and Zn (Cd-free), whose disadvantage is the high potential of evaporation during soldering. For this reason, in the most recent filler materials, Zn is replaced by Cu, Ni and Mn, sometimes accompanied by small amounts of indium or tin to decrease the melting temperature.

The brazing of Al_2O_3 components using metallic filler metal can be performed with satisfactory results by using the furnace brazing technique. If the filler materials for brazing and ceramic surfaces are not properly cleaned and processed (degreasing, etching, polishing), there can appear oxide layers that diminish the resistance to shear.

In the case of Al_2O_3 components, the combinations of filler materials that provide satisfactory results are $Ni_{59}Cr_{13.5}Zr_{4.3}Fe_{4.2}Nb_{2.9}Si_{15.3}B$ and $Fe_{73.5}Cu_1Nb_3Si_{15.5}B$. Metallographic examinations revealed the occurrence of oxide layers at the interface between the ceramic and filler materials, containing intermetallic compounds that ensure connection continuity but do not provide sufficient mechanical strength. For this reason, a protective atmosphere must be used during the brazing of reactive ceramic materials.

Acknowledgements

The research work was financially supported by the Romanian National Program for Research within the framework of the projects: 'Materials and performance technologies designed to obtain asphalt milling knives' no. PCCA 188/2012—MATFREZ and 'Nanomaterials filler materials for joining ceramics', no. 71-118/2007—NANOCERAD. In addition, the author would like to thank her colleagues Prof. PhD Eng. Victor Geanta from Politehnica University of Bucharest and PhD Eng. Horia Binchiciu from SC. SUDOTIM AS Timisoara (Romania), for relevant advice and active collaboration during the experimental research.

Author details

Ionelia Voiculescu

Address all correspondence to: ioneliav@yahoo.co.uk

Politehnica University of Bucharest, Romania

References

[1] Davis JR. Phase diagrams ASM Specialty Handbook: Copper and Copper Alloys. ASM International; 2001. pp. 369-382. ISBN: 978-0-87170-726-0

[2] Huhne T, Hofmann E, Fussel U. Soldering and brazing of copper and copper alloys. Deutches Kupferinstitut.de. Broshure of Deutsches Kupferinstitut Berufsverband e. V. Am Bonneshof 5 40474 Düsseldorf Germany: 2015;4-57.

[3] Voiculescu I, Geanta V, Vasile IM, Binchiciu EF, Winestoock R. Chemical elements diffusion in the stainless steel components brazed with Cu-Ag alloy. In: International Conference on Innovative Research 2016-ICIR Euroinvent 2016, IOP Conference Series: Materials Science and Engineering. 2016;**133**:1-8. DOI: 10.1088/1757-899X/133/1/012016

[4] Binchiciu E, Fleser T, Voiculescu I. Chemical active coated rods used for brazing. Advanced Materials Research. 2015;**1111**:25-30

[5] Binchiciu E, Fleser T, Voiculescu I, Binchiciu A. Coated rods for brazing structures used in high safety conditions. Advanced Materials Research. 2015;**1128**:224-229

[6] Binchiciu E, Fleser T, Voiculescu I, Tihanov Tanasache D, Dragomir A. Composite materials for mechanized brazing. Advanced Materials Research. 2016;**1138**:119-124. ISSN: 1662-898

[7] Voiculescu I, Binchiciu H, Geanta V, Iovanas DM. Research project 'Materials and performant technologies designed to realize knives for milling the asphalt" no. PCCA 188/2012 – MATFREZ, 2014

[8] Xing F, Qiu X-M, LI Y-D. Effects of Sn element on microstructure and properties of Zn–Cu–Bi–Sn high-temperature solder. Transactions of Nonferrous Metals Society of China. 2015;**25:**879-884

[9] Matthey J. Metal Joining, Silver Brazing Alloys and Fluxes [Internet]. pp. 4-39. Available from: http://www.jm-metaljoining.com

[10] Davis JR. Stainless steel cladding and weld overlays. In: ASM Speciality Handbook: Stainless Steels. ASM International; 1994. pp. 109-110

[11] Willingham J. Filler metals and fluxes for brazing—tungsten carbide. In: Matthey Johnson editor. Metal Joining. 2016, pp. 1-31. Available from: http://www.jm-metaljoining.com

[12] Weil KS, Hardy JS, Rice JP, Kim JY. Brazing as a means of sealing ceramic membranes for use in advanced coal gasification processes. Fuel. 2006;**85**:156-162

[13] Suganuma K, Okamoto T, Koizumi M, Fujita T, Niihara K. Joining of silicon nitride with metallic interlayers. Journal of the American Ceramic Society. 1990;**73**(5):1221-1227

[14] Olesinski RW, Abbaschian GJ. The Cu-Si (copper-silicon) system. Bulletin of Alloy Phase Diagrams. 1986;**7**(2):170-178

[15] J. T. Klomp, Y. Ishida, Interface Chemistry and Structure of Metal-Ceramic Interfaces, Fundamentals of diffusion bonding. Elsevier Science Publishers, Amsterdam, Oxford, New York, Tokyo, 1987: **43**

[16] Binchiciu A, Voiculescu I, Geantă V., Binchiciu H, Ştefănoiu R, Negriu RM, Binchiciu E. Ecological coated core wires of silver alloys for brazing, is made by extrusion coating with deoxidizing flux with metal powder addition. [Derwent Innovation Index ISI-Web of Science/Science Citation Index Expanded]. Patent Number: RO125835/30.07.2012

[17] Binchiciu A, Voiculescu I, Geantă V, Binchiciu H, Ştefănoiu R, Binchiciu E. Coated silver alloy rods of reduced hygroscopicity for brazing, comprises silver and tin deposited by extrusion deoxidizing flux made of synthetic vitreous precursors and metal powder additions. [Derwent Innovation Index ISI-Web of Science/Science Citation Index Expanded]. Patent Number: RO125836/18.03.2010

[18] Schwartz M. Ceramic Joining. Materials Park, OH: ASM International; 1990. p. 166

[19] Klomp JT. Interface Chemistry and structure of metal-ceramic interfaces. In: Nicholas MG, editor. Joining of Ceramics. London: Chapman and Hall; 1990

[20] Voiculescu I, Geanta V, Stefanoiu R, Binchiciu H, Daisa DD. Alumina brazed using metallic alloys. In: International Conference on Innovative Technologies. Rijeka: In-Tech; 2012. pp. 523-527

[21] Voiculescu I, Geanta V, Daisa D, Alexandrescu E. Research project Nanomaterials filler materials for joining ceramics – NANOCERAD, no. 71-118/2007

Recent Advances in Solderability of Ceramic and Metallic Materials with Application of Active Solders and Power Ultrasound

Roman Koleňák

Additional information is available at the end of the chapter

http://dx.doi.org/10.5772/intechopen.69552

Abstract

The Chapter deals with solderability of ceramic materials by ultrasound and suitable selection of soldering alloy. The solderability issue of ceramic materials consists mainly in the fact that the ceramic materials are not wettable by the common solders, due to their ionic and covalent bond between the atoms. However, there exist several ways to ensure the wettability of ceramic material surface. One of them is for example coating of ceramic material by a metallic layer. Anyway, a more perspective solution seems to be the application of soldering alloys which are alloyed with a small amount of alloying elements which exert high affinity to some component of ceramic material. The basic group of such solders are the so-called active solders containing from 1 to 5 wt. % Ti. Another group of solders, which may wet the ceramic material, are the solders alloyed with a small amount of lanthanides, for example La, Ce etc. The content of lanthanides varies from 0.5 up to 2 wt. %. The last group consists of the solders containing indium in the amount from 20 to 100 wt. %. The aim of study was to compare these three groups of soldering alloys from the viewpoint of mechanism of bond formation. The interactions between the solder and ceramic substrate were analysed and shear strength of fabricated soldered joint was determined. To improve the solderability, the soldering in combination with an active power ultrasound was also employed.

Keywords: solder, active solder, metallic materials, ceramic materials, ultrasonic soldering, wetting, shear strength, tin, indium, lanthanum, EDX analysis

1. Introduction

Creation of a sound bond of ceramic materials mutually or in combination with metal depends on the fact that whether a close contact to atomic distance is formed on the but surface and whether ceramic material is wetted with the metallic solder. Wetting the ceramics with a metallic solder is an essential precondition for soldered joint formation.

Several ways may be used for ensuring the wettability of a ceramic material. For example, the deposition of solderable coating on ceramic material is often employed.

However, more modern methods seem to be the application of solders which are alloyed with some active metal such as Ti, In, Si, Al, Mg and/or lanthanides, which exert a strong chemical affinity to oxygen.

The first group comprises the active solders based on Sn or Sn-Ag, which may contain from 1.5 to 4 wt.% Ti. Another group of solders, which may wet the ceramic material, consists of solders alloyed with a small amount of lanthanides, for example, La, Ce and so on. The content of lanthanides usually varies from 0.5 to 2 wt.%. The last group are the solders containing indium in the amount from 20 to 100 wt.%. However, solving the mentioned issue still involves the requirement to ensure a joint that should resist the effects of residual stresses formed owing to different coefficients of thermal expansivity of metal and ceramics, which usually exerts much lower expansivity. This condition is most essential from the viewpoint of ensuring the reliability of ceramics/metal joints [1].

2. Methods of soldering ceramic materials

The biggest issue in the fabrication of joints of ceramic and non-metallic materials with metals consists in the fact that the commercial solders generally do not wet the ceramic materials. Poor wettability of ceramics is caused by the non-metallic character of bond, which is mostly ionic or covalent, in contrary to metallic materials with different structure of energy levels of electrons.

The following methods are mostly used to ensure the wettability:

- Metallizing of ceramics (e.g. metallic deposit of Mo-Mn paste, physical vapour deposition (PVD) and chemical vapour deposition (CVD) processes).
- Application of an active brazing alloy and/or solder alloyed with active elements (Ti, Zr, Hf).

All the mentioned technologies are applicable for soldering the ceramics/metal combinations. Regarding the versatility of technology, its ease end economic efficiency, there is increasing tendency to apply the soldering with an active solder. Soldering of metallized ceramics requires several additional operations, when compared to the application of active solders [2].

2.1. Soldering of metallized ceramics

Metallizing of ceramics eliminates the issues connected with its wettability. Regarding the selection of metallizing, it is necessary to know at what working temperature the soldered part will be servicing. Then, either solder and/or brazing alloy will be used. The desirable metallic layer may be obtained:

- By burning-in of metallic solution either of heat-resistant metals such as Mo, Mn, W or precious metals such as Ag, Au, Pt and so on.

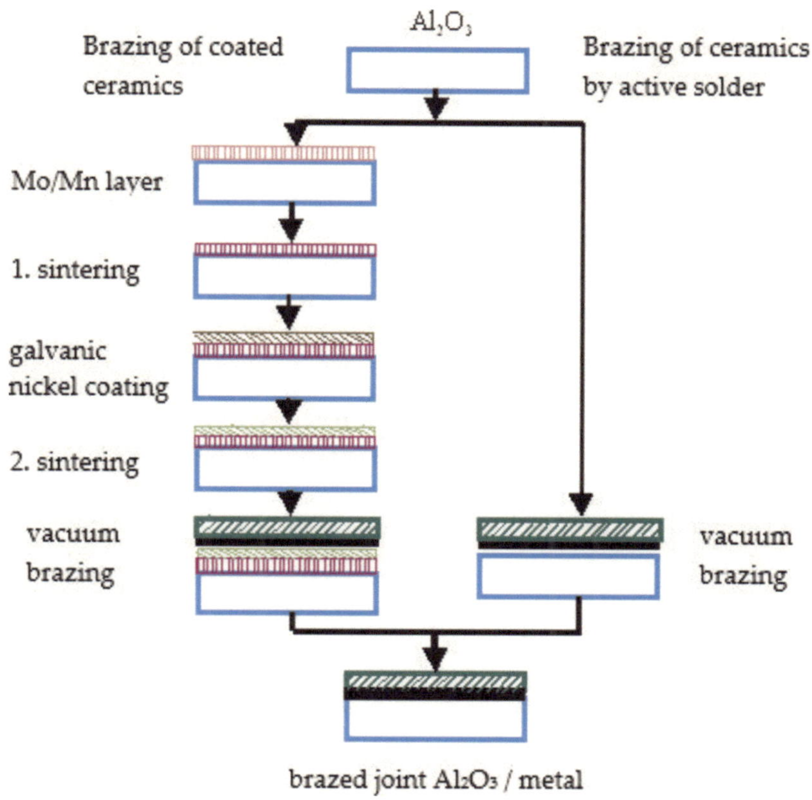

Figure 1. Schematic representation of vacuum brazing of metal and ceramics, metallized (with Ag72Cu brazing alloy) in comparison with non-metallized substrate (brazing with active Ag72CuTi brazing alloy) [3].

- By physical and chemical deposition, with a thin coatings, for example, of Au, Ag and Ni are created.

The classical method of metallizing Al_2O_3 ceramics is shown in **Figure 1**. A powder deposit (70% Mo, 17.5% Mn, 3.5% titanium hydride, 9% kaolin) is applied onto ceramics surface, followed by sintering at 1200–1400°C/10 min in humid hydrogen. If a silver brazing alloy (e.g. Ag72Cu) is used for brazing, then the sintered surface must be coated with nickel, to allow the brazing operation in vacuum. Optimum thickness of metallizing is 0.02 mm. Deviations from this value may affect the strength of soldered/brazed joint. The paste is deposited on the soldered surface by a brush and on/or by silk-screen printing. The layers formed by this process ensure at certain conditions a good reliability, however, paid by high power demands and necessity of costly equipment and extensive safety precautions at their preparation [4].

Chemical reactions between the Al_2O_3 substrate and molybdenum-manganese suspension allow joint formation. In case of this method, the reactions take place in the following manner:

- At sintering the metallic suspension with Mn in humid hydrogen, the manganese oxide is formed by the reaction [5]:

$$Mn + H_2O \rightarrow MnO + H_2 \quad (1)$$

- The formed oxide then react with Al_2O_3, forming thus the manganese-aluminium spinel as the product of reaction [5]:

$$MnO + Al_2O_3 \rightarrow MnAl_2O_4 \tag{2}$$

2.2. Solderable metal coatings on ceramic materials formed by physical and chemical deposition

The methods for the formation of metallic layers and ceramic or glass substrates may be divided into several groups. The layers are obtained by chemical deposition from solutions of appropriate metal salts (e.g. Ni, Ag, etc.) by the aid of chemical agents, such as formaldehyde, hypophosphite, glucose and so on, with eventual heat treatment of the segregated coating. These layers are used in cases when small areas are metallized, eventually a motive composed of individual narrow conductive tracks, which are not mechanically loaded. Soldering of such layers requires the application of solders containing Sn, In, Cd and so on, with a low melting point (max. 200°C) and a minimum dwell time on soldering temperature [6].

Other known methods comprise the physical deposition by vacuum evaporation and sputtering of metallic layers as shown in **Figure 2**. These processes employ a high vacuum and temperatures within the range from 200 to 500°C. In case of these methods, the coating particles are deposited rather in physical manner, not by the aid of chemical reactions, as in the case of CVD (chemical vapour deposition) processes.

The thickness of evaporated or sputtered layers varies in the order of nm to μm. To prevent the de-alloying (de-wetting) of the metallic layer to the molten solder, the multiple deposition of system consisting of several metals, for example, Cr/Ni-Ag, Ni-Ag and Cr- Ni/V-Ag, is used.

To ensure a good spreadability of solder over the entire surface, the final layer is often formed by evaporation or sputtering of Au or Ag, which is dissolved in the used solder and will thus create suitable conditions for a good wettability of entire surface by the solder. Chemical reaction of the deposited layer with the material of ceramic substrate does not occur in both mentioned deposition methods. The bond between the layer and ceramics is mostly of adhesion character [2]. The solderable coatings for ensuring the wettability of poorly wettable surfaces are mostly employed in the electrotechnics and electronics. The advantage of well-solderable metal coatings consists in the fact that the soldering process can be accomplished in very short soldering times. The used coatings are either meltable or soluble (**Figure 3**). The soluble

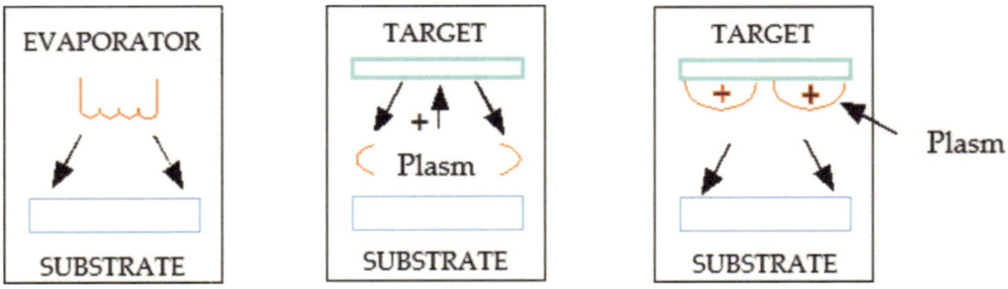

Figure 2. Main physical deposition methods applied for metallizing of ceramics [7]: a, evaporation; b, diode sputtering; c, magnetron sputtering.

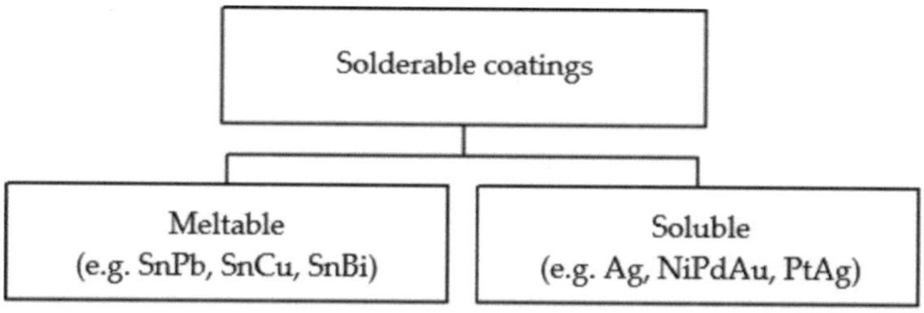

Figure 3. Basic division of solderable coatings [7].

coating may consist of one metal or of several metal layers deposited subsequently on the substrate. The meltable coating is formed by the deposition of the lead- and/or lead-free layer of soldering alloy on the substrate surface.

The system of dissolvable solderable coating in electronics mostly consists of an adhesion layer (Cr, Ni, Ti, Al), a diffusion barrier (Ni, Cr) and a solderable layer (Ag, Au, Cu, Ni), whereas its composition depends on the substrate material and parameters and conditions of soldering. For preserving a good solderability, for example tinning with pure tin, metallizing with Sn-Pb solders, gold and/or silver plating used to be applied until now. However, the application of lead-free soldering necessitates the new approaches also in the field of surface finishing, especially in printed circuits and also outlets of electronic components. The meltable coatings must be selected without lead content, regarding thus the environmental viewpoint. The coatings composed of SnBi and SnCu, eventually coating of pure tin, start to be more favourable nowadays [8].

A coating composed of a system of Cr/Ni-7%V/Ag layers, prepared by PVD sputtering, was approved experimentally. The chromium layer has the adhesion function, Ni-V is the diffusion barrier and Ag layer ensures a good wettability and spreadability in a short time as shown in **Figure 4** [8].

The SnIn52 solder with a melting point of 120°C was proposed for soldering. The soldering temperature was selected just by 10°C higher than its melting point with a minimum dwell time, needed for a proper spreading of the solder.

2.3. Soldering of ceramic materials with an active solder

The wettability of ceramic material may be improved by reducing the interphase stress on the ceramics/substrate interface, by solder alloyed with an active element (mainly Ti, Zr or Hf is mostly mentioned in the literature) with a high affinity to oxygen, which reacts with ceramics during the soldering process, whereby bonds on the interatomic level are formed. The basic chemical reaction between Ti and oxidic ceramics has a general form [16]:

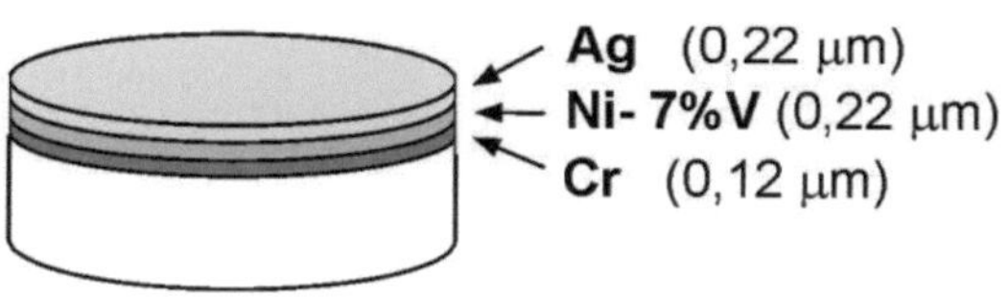

Figure 4. Distribution of metallic layers on a substrate [8].

$$y/Ti/+MxOy \rightarrow yTiOi + x/M/ \quad (//- \text{melt}) \tag{3}$$

Titanium can bond considerable amount of oxygen in ceramics of oxidic type. Therefore, several oxides can be formed, for example, between the AgCuTi solder and Al_2O_3 ceramics: TiO, Ti_2O_3, Ti_3O_5, Ti_4O_7 and TiO_2. The reaction product formed on the contact area at individual types of oxidic and non-oxidic ceramics is several μm in thickness and depends on the soldering conditions and solder type. The reaction product alters the surface energy of ceramics and allows its wetting by the solder [17]. Higher concentration of active element may in some cases increase the joint brittleness; therefore, its amount must be limited. The foreign sources refer the maximum limit to 4 wt. % Ti in an eutectic solder-type Ag72Cu. Chemical changes occurring on the ceramics-active solder interface are very complex since the concentration gradients are formed [1].

The process of soldering ceramics with metal is significantly simplified by the application of active solders. Soldering ceramics with an active solder is also more economically efficient, since the multi-stage metallizing processes with demanding inter-operational annealing are unnecessary [9].

2.4. Division of active solders

The active solders, similar as the commercial solders, are classified by the melting temperature to solders, brazing alloys and high-temperature solders. The difference is seen just in the case of active solders, which may be further divided into high-temperature solders and the mechanically activated solders. Due to reaction capability of an active element, the soldering temperature for high-temperature activated active solders must be higher than 780°C (at the application of active Ti). Nevertheless, the active solders are molten at the temperature around 220°C. The division of active solders and their chemical base is given in **Table 1**.

2.5. Active solders

The brittle materials as vitreous glass (SiO_2), silicon, graphite and so on can be soldered with active solders. Base metal of active solders is mostly tin, lead or indium and the alloys created

Active solders	Application temperature	Chemical composition
Active solders	From cryogenic temperatures up to about 150°C	Sn, In and Pb based, for high-temperature activation (e.g. Sn95Ag5Ti3) Sn and In based, for mechanical activation (e.g. SnAg6Ti4Ce)
Active brazing alloys	up to 350°C	Ag, Cu and Au based (Ag72CuTi1,5)
High-temperature active solders	up to 900°C	Ni, Co, Pd and Pt based (e.g. Ni70Hf30)

Table 1. Division of active solders by the melting temperature [10].

on their basis. However, the lead-containing solders are not suitable for soldering in vacuum, since considerable evaporation of lead and also furnace contamination occurs. When soldering with lead, a through-flow atmosphere with pure argon, eventually helium with overpressure attaining 0.1–0.2 MPa, should be used [11].

The solders used in joint assembly are capable to compensate the stresses resulting from different thermal expansivity, owing to their plastic straining by the shear or creep mechanism. In this way, the most significant reduction of residual stresses at preserved joint simplicity may be achieved. The presence of an active element ensures a good wettability of soldered parts.

The active solders can be used for soldering unusual combinations of metallic materials (e.g. CrNi steel, Mo, W, Ti, Cr, etc.) and non-metallic materials, mostly of brittle character (vitreous glass, sapphire, carbon, silicon and also almost all types of ceramics) [11]. Such joints are applied mainly in electronics and electrotechnics, where lower strength and thermal resistance of the joint are sufficient. For example, the soldering of glass in lasers and spectroscopes, creation of electric connections with graphite, connecting the heat exchangers to ceramic electronic substrate of Al_2O_3, or AlN [12]. The solders also allow fabricating the vacuum-tight joints in vacuum and cryogenic technology, where the application of indium-based solder was well approved [11].

Great attention is paid to issues of soldering with active solders, especially for the electronic applications. However, there are still many issues, regarding the achievement of the desired utility properties of joints, to be solved.

As already mentioned, the active solders may be further divided according to the way of active element activation in the solder to the solders destined for:

- high-temperature activation,
- mechanical activation.

The active brazing alloys for high-temperature activation based on Ag and Ag-Cu, designated as CB 2, CB 4, CB 5 and CB 6, are supplied by Umicore-BrazeTec GmbH, Germany. The brazing alloy designated as CB 10, with composition AgCu25Ti10, and the alloy designated as CB 11, with composition AgTi10, are supplied in the form of brazing paste. The working temperature of these brazing alloys is 850–1050°C. The solders for ultrasonic activation, designated as CERASOLZER, are supplied by the Japanese company KURODA ELECTRIC.

The solders for mechanical (ultrasonic) activation, designated as S-Bond, are supplied by the Euromat, GmbH, Germany. It concerns, for example, the solders type S-bond 140 (based on Bi-Sn-Ag-Ti), S-bond 220 (based on Sn-Ag-Ti), S-bond 400 (based on Zn-Al-Ag) and so on. The soldering temperature for the S-bond 140 solder is from 150 to 160°C, for the S-bond 220 solder it is 240–260°C and for the S-bond 400 solder the soldering temperature is 420–430°C. It was proved experimentally, that the S-bond 220 (SnAg6Ti4Ce) solder has wetted the Al_2O_3 ceramics with the wetting angle of 62°, when applied at a temperature of 860°C in vacuum of 10^{-2} Pa. This experiment was performed in cooperation with the Institute of Materials and Machine Mechanics at the Slovak Academy of Sciences in Bratislava.

2.6. High-temperature activation

It is a process being performed at high temperature (850–950°C), mostly in vacuum furnace with the shielding atmosphere of argon as shown in **Figure 5**. It was found out experimentally that at activation with Ti-active element, the lowest temperature at which wetting of Al_2O_3 ceramics may be achieved is 780°C. However, wetting depends also on the content of Ti-active element in the solder [7].

With increasing soldering temperature (activation temperature), the wetting of ceramic material is enhanced by high-temperature activation but, in contrary, if a ceramic material is soldered in combination with metallic material, the degradation of base metal by erosion may occur.

2.7. Mechanical activation

The mechanically activated solders allow wetting the metallic and non-metallic materials at considerably lower temperature than in the case of high-temperature activation. Moreover, there is no need to solder in vacuum and/or in the shielding atmosphere. In dependence on soldering alloy type, this temperature may vary from 150 to 430°C, thus in the temperature range of soldering with solders, which are based on Bi-Sn, Sn-Ag or Zn-Al, as already mentioned. This unique capability is allowed by a slight addition of the following metals: lanthanum, cerium, yttrium and samarium (lanthanides), which occur in the soldering alloy matrix. These metals at the same time create a protective barrier for the active metal (Ti) as shown in **Figure 6** [13]. Mechanical activation can be performed by

- intrusions
- vibrations (50–60 Hz)
- ultrasound (20 60 kHz)
- friction (different processes)—**Figure 7**.

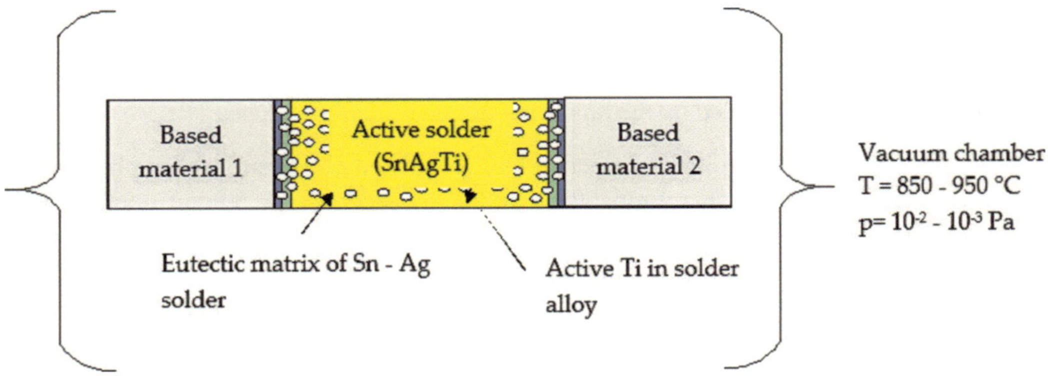

Figure 5. Scheme of soldering with high-temperature activation in vacuum [14].

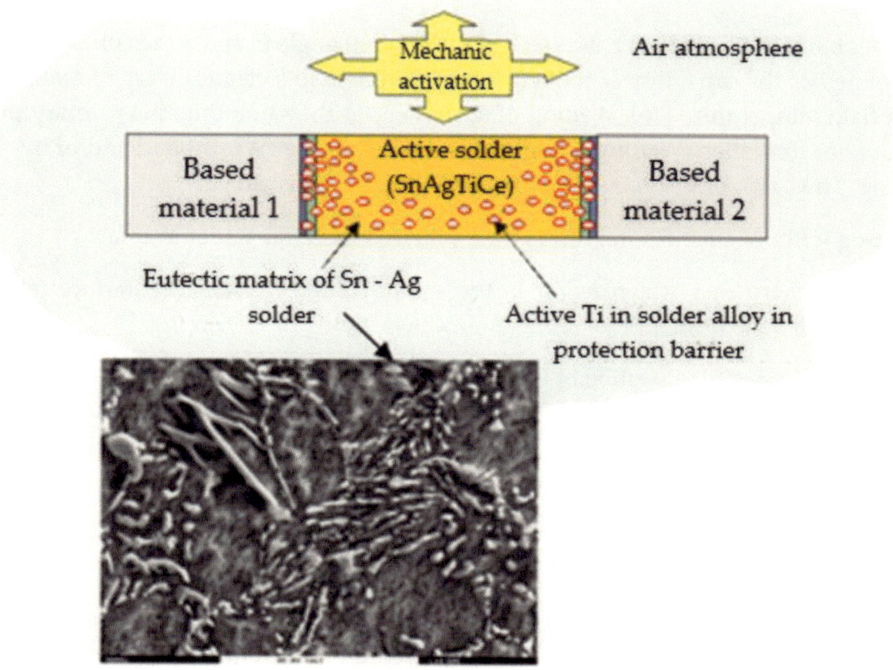

Figure 6. Scheme of soldering with mechanical activation in air without flux—S-Bond process [14].

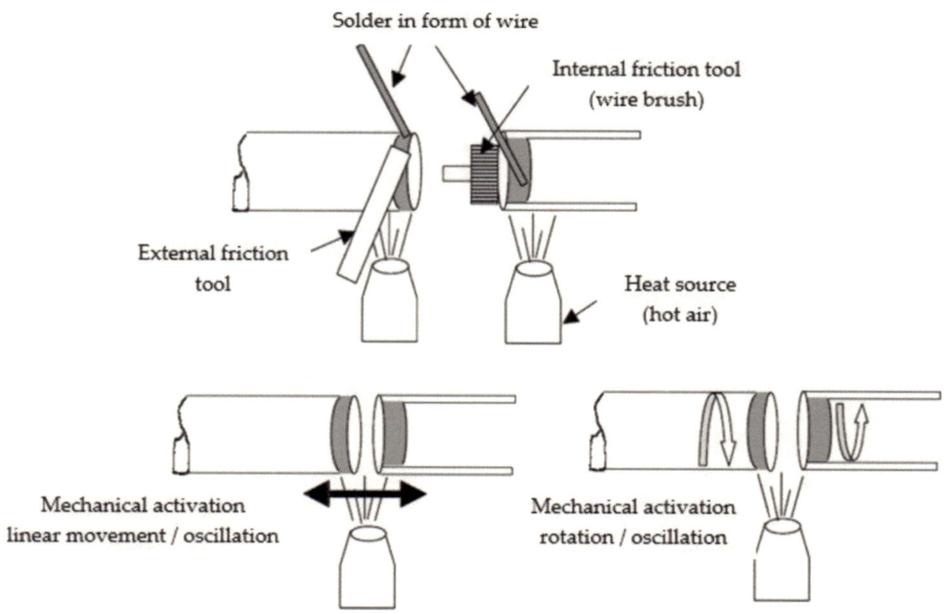

Figure 7. Scheme of different ways of mechanical activation of solder by friction [14].

2.8. Wetting

The wettability is qualitatively assessed by the wetting angle Θ and adhesion work Wad [15]. It is defined as the capability of molten solder to adhere to a clean surface of material joined at a certain temperature [16]. Wetting of surfaces is at the same time the primary precondition for joint formation. Wetting is realized either by Van der Waals bond and/or by chemical bonding. Thus, two basic types of wetting may be distinguished [15]:

- Wetting without chemical reaction in the soldered material-solder interface.
- Wetting, where chemical reaction in the solder-soldered material interface takes place, whereby also the formation of reaction products (new phases) occurs.

The equations describing wetting of liquid droplet on a solid material surface were derived on the basis of physical and chemical principles. The droplet during wetting takes such a shape, at which the energy of solder-material-atmosphere (vacuum) system is minimum, and thus the interatomic forces may be exerted [16]. The basic Young's equation for a weak reaction system may then result from the facts mentioned [17]:

$$\gamma SV - \gamma SL = \gamma LV \cos\Theta \tag{4}$$

where γ_{SV} is the surface energy between the material surface and atmosphere, γ_{SL} is the surface energy between the material surface and solder, γ_{LV} is the surface energy between the solder and atmosphere and Θ is the wetting angle.

The magnitude of wetting angle is the qualitative criterion of wetting. **Figure 8** shows two basic cases which may occur at material wetting with a liquid solder.

As obvious from **Figure 8**, the solder does not wet the material for which the value of wetting angle $\Theta > 90°$ ($\gamma_{SL} < \gamma_{SV}$). If the value $\Theta < 90°$, the solder wets the material. By the equation of wetting (4), the driving force of process is the ($\gamma_{SV} < \gamma_{SL}$) difference. By the magnitude of wetting angle, we distinguish different degrees of wetting—**Table 2** [16].

Lowering the contact angle of wetting and increasing the value of adhesive work at a defined solid substrate may be affected by the selection of a suitable solder. However, for wetting

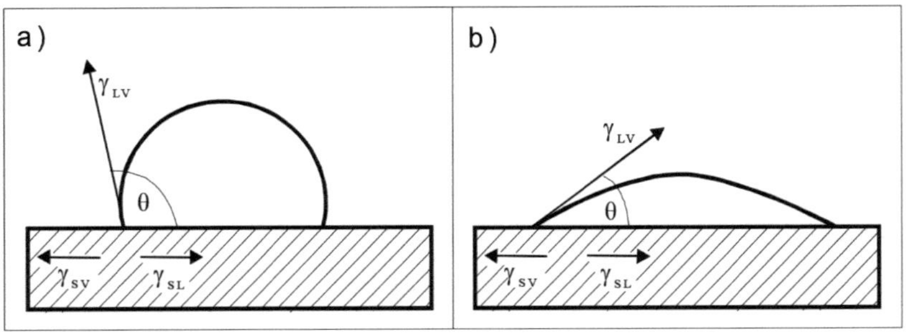

Figure 8. Scheme of wetting for a weak reaction system. (a) Solder does not wet the material; (b) solder wets the material.

http://dx.doi.org/10.5772/intechopen.69552

Wetting degrees	Contact angle
Excellent (full)	$\Theta = 0°$
Very good	$\Theta < 20°$
Good	$20° < \Theta < 90°$
Poor	$\Theta > 90°$

Table 2. Assessment of wetting.

ceramic materials, it is necessary to use an active solder, that is, the solder containing the element chemically active towards ceramics (Ti, Zr or Hf). Such an element is chemically bond with some component of ceramics, so improving its wettability and thus also the joint strength [15]. Scheme of wetting for the system with reaction on the interface is shown in **Figure 9**. In such a case, Young's equation according to Ref. [18] attains the following form:

$$[\,\gamma SV - (\gamma RL + \gamma SR + \Delta G\,.\,t/\Omega\,)\,] = \gamma LV \bullet \cos\Theta \qquad (5)$$

where γ_{SV} is the surface energy between the material surface and atmosphere, γ_{LV} is the surface energy between the solder and atmosphere, γ_{RL} is the surface energy between the solder and reaction layer, γ_{SR} is the surface energy between the reaction layer and material, Θ is the wetting angle, Ω is the gram atomic volume, t is the thickness of reaction layer and ΔG is the change in free energy per one mole of reaction product.

2.9. Compensation of different thermal expansivity of the ceramics/metal-soldered joints

The residual stresses in soldered joint of two materials with different coefficients of thermal expansivity belong to one of the most serious issues for ensuring the reliable joints of ceramics/metal. Ceramic materials exert in general lower coefficients of thermal expansivity and are not capable of plastic strain. On the contrary, the metals exert much higher coefficients of thermal expansivity and are in certain extent capable of plastic strain. These are the main reasons for the formation of residual stressed during cooling down of ceramics/metal joints, which must be compensated.

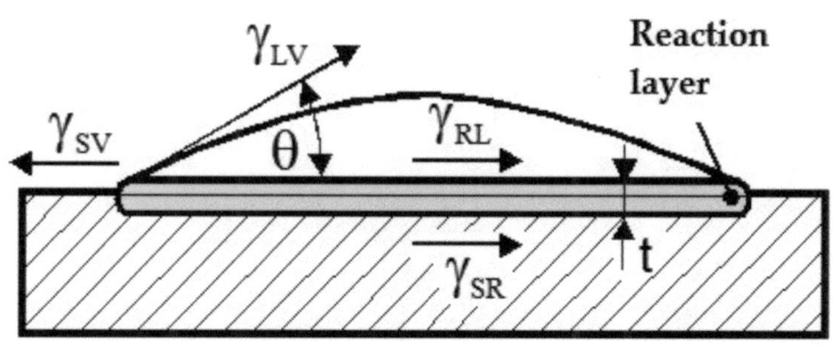

Figure 9. Scheme of wetting for the system with reaction on the interface [18].

Reducing the level of residual stresses may be more or less attained in the following ways [7]:

- *By the selection of suitable material couples.* Application of joined materials with similar coefficients of thermal expansivity is not a practical solution in most cases, since the joined materials are mostly selected on the basis of other desired properties and not on the basis of their thermal expansivity.
- *Soldering temperature.* By selection of a suitable solder (regarding thermal resistance of the joint), reduced soldering temperature may be achieved.
- *Shape of joint and thickness of the materials joined.* A simple type of joint should be selected, attaining thus the compressive loading in ceramic material. It should be taken into account that the ceramic materials do not exert constant value of strength limit. This is in practice exerted in such a way that during loading of ceramics a crack may be formed without any strain.
- *Size of clearance between the soldered parts.* In soldering-combined materials, it is required that the joint clearance should be generally greater than in the case when soldering the same materials (**Figure 10**). It is recommended to select the clearance value within $s = 0.2–0.5$ mm [19].
- *Application of interlayer.* If allowed by the design solution of the joint, the reduction of residual stresses can be attained also by the application of an interlayer, where material is selected on the basis of elastic, plastic and thermal properties. At the application of interlayer with a small coefficient of thermal expansivity, close to that of ceramics (W, Mo), lower stresses are formed in the ceramics. In the case of application of interlayer with a low yield point (Ni, Cu, Al), the level of stresses is reduced due to their relaxation by the slip mechanism. At the application of a composite interlayer (e.g. cermet-sintered interlayer composed of ceramic and metallic powders) with gradual transition from metal expansivity to expansivity of ceramics, the creation of gradient of physical properties is considered as shown in **Figure 11**. In interlayer selection, it is also necessary to consider the interactions with the molten solder: segregation, formation of brittle structures on the interface and dilution in the solder.
- *Application of solder.* At the application of metal solders in the joint assembly, these are capable to compensate the stresses resulting from different thermal expansivity by their plastic straining via the slip or creep mechanism. In this way, the most significant reduction of residual stresses at preserved joint simplicity may be attained [2].

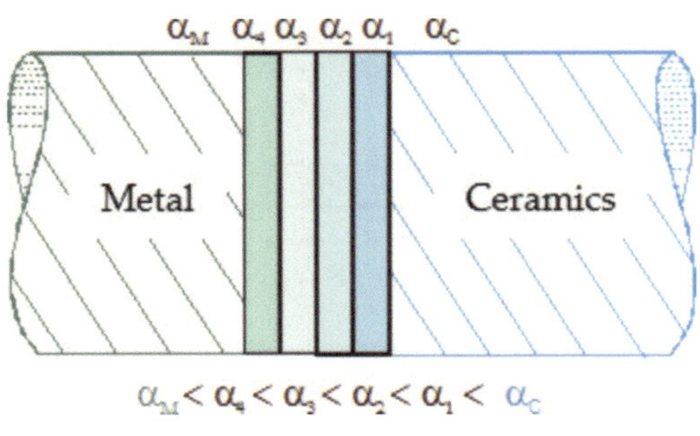

Figure 10. Complex interlayer [20].

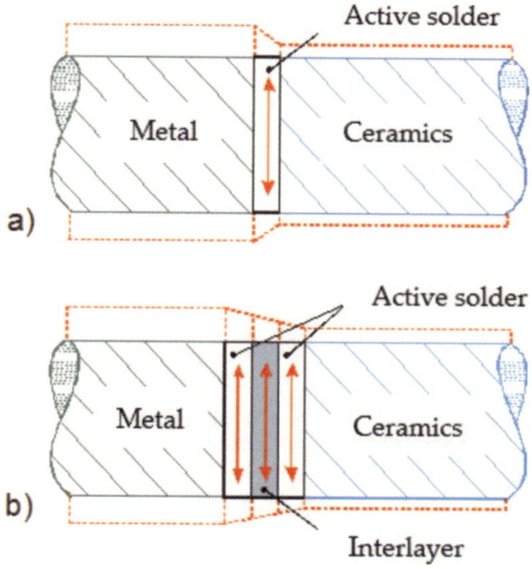

Figure 11. Reduction of residual stresses in contact zone by application of plastic metallic interlayer: a, joint without interlayer; b, joint with interlayer [21].

2.10. Indium-based solders

The indium-based solders, for example, type 100In or the solders type In-Sn with a higher indium content, are characteristic with their unique soldering properties. These solders allow to wet a wide range of metallic, non-metallic and ceramic materials, for example non-metallics such as glass, glazed ceramics, mica, mullite, quartz, fibre optic glass, lead glass, liquid crystal glass, metallized glass, optical glass, pyrex, quartz glass, sapphire, silica, silica glass, soda lime glass, and various metallic oxides [22]. During soldering process performed with high-indium solders in air, the indium suboxides are formed, which react with the surface oxides on the soldered substrate at the formation of a strong bond between the substrate and solder. An example of soldering the ITO ceramics (In_2O_3/SnO_2) presents the study [23], which deals with the application of soldering alloy type Sn-In-Ag-Ti at ultrasonic soldering of ITO ceramics. It was found that the Sn-In-Ag-Ti solder reacts with the surface of ITO substrate, whereby the wettability of materials joined and bond formation was ensured.

The goal of the next research within the study [24] consisted of soldering metallic (Cu, Ni, Al, Ti, AISI 316 steel) and ceramic materials (Al_2O_3, SiC) by the aid of solders with a high content of indium at power ultrasound application. The solders with composition 100In (5N purity) and 70In30Sn (4N5 purity) were used in experiments. The UT equipment with a frequency of 40 kHz and an output power of 400 W with 2 μm amplitude was used for experiments. The scheme and description of this equipment is shown in **Figure 12**. Soldering temperature was 20°C above the liquidus of the solder tested. The dwell time at soldering temperature was 30 s and the time of ultrasound action on the soldered joint was 5 s. Heating of specimens was ensured by the hot-plate method with temperature control via a thermocouple type NiCr/NiSi. The shear strength of metallic and ceramic substrates was assessed. The test specimens of substrates were prepared in the form of disks with a diameter of Ø 15 mm and

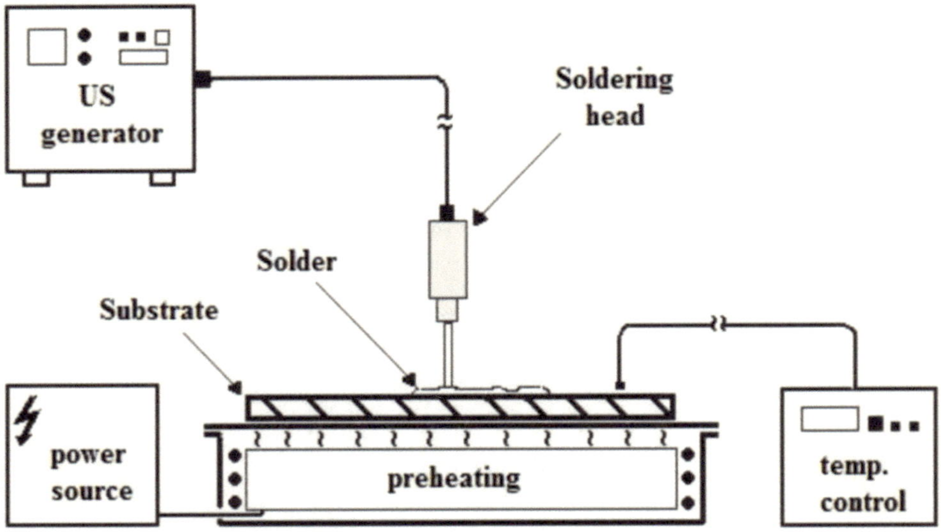

Figure 12. Scheme of ultrasonic device used for soldering.

1.5–2 mm in thickness. The test specimen is shown in **Figure 13**. The procedure of specimen preparation is shown in **Figure 14**. The shear gap was selected to 0.1 mm, which corresponds to 2% from the sheared diameter of the roll formed of the solder. Uniform testing rate was 0.5 mm/min.

The 100In and 70In30Sn solders with a high indium content wetted all studied metallic (Cu, Al, Ni, Ti and AISI 316 steel) and ceramic materials (SiC and Al_2O_3) at the application of power ultrasound with the frequency of 40 kHz. For comparison, the 100Sn solder wetted all metallic materials but it did not wet the ceramic materials in spite of ultrasound assistance. The 100Sn solder cannot be used for soldering ceramic and non-metallic materials.

The results of mechanical tests of soldered joints fabricated by the use of 100In solder are given in **Figure 15**. The highest shear strength of 12.5–54 MPa is achieved at the soldering of metals. The lowest shear strength was observed with aluminium and the highest with copper. Considerably lower shear strength values are achieved at the application of 100In solder on the ceramic materials, varying from 3.5 to 6 MPa. Higher strength was attained with Al_2O_3 ceramics and lower with SiC ceramics.

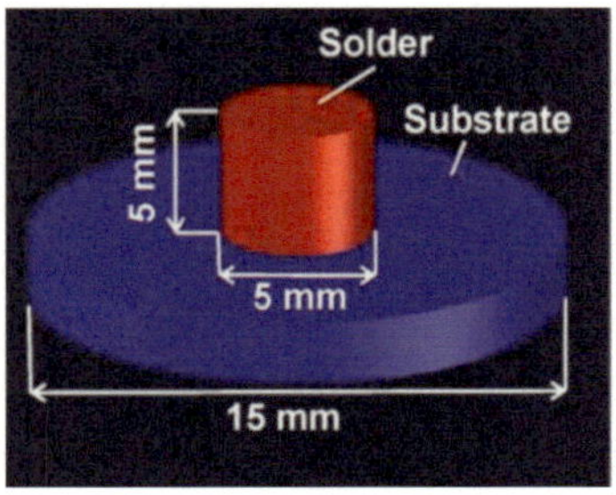

Figure 13. Fabricated test specimen.

http://dx.doi.org/10.5772/intechopen.69552

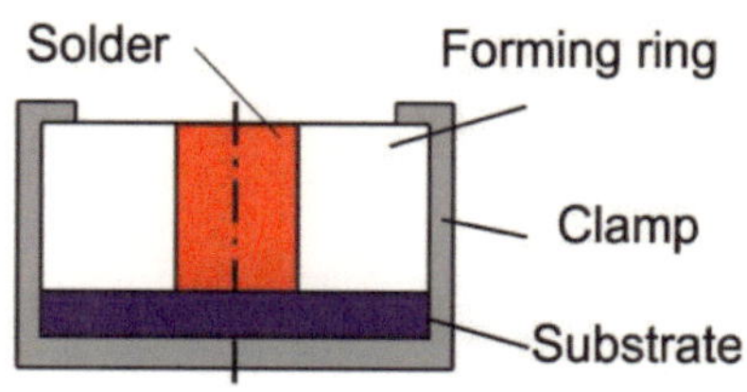

Figure 14. Specimen layout at its preparation for the shear test.

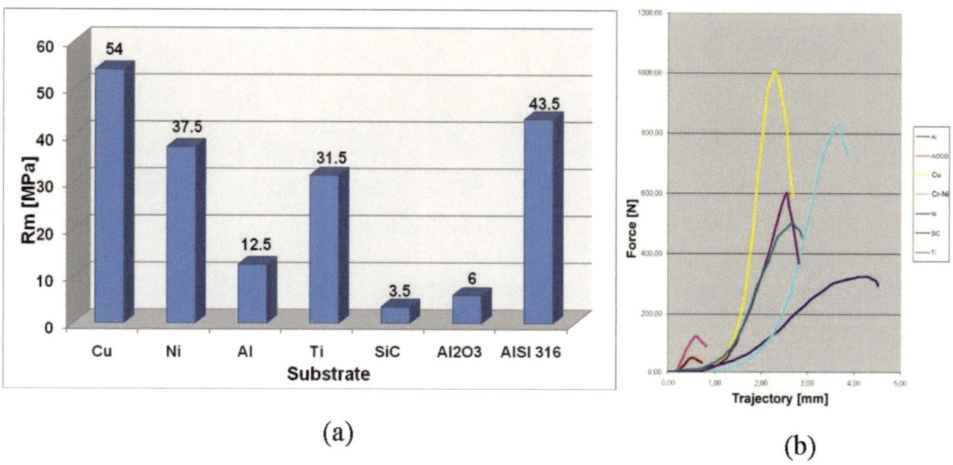

Figure 15. Shear strength of joints fabricated with solder type 100In.

At comparison of the results of shear strength attained with 100In and 70In30Sn solders (**Figure 16**), it was found that with the solder containing tin higher strength values are achieved, both on the metallic and ceramic materials. This is caused by the fact that the matrix of 70In30Sn solder exerts the eutectic structure [25, 26]; therefore, it offers also higher mechanical resistance, when compared to pure indium solder. At the In content of 70 wt. %, the solder preserves also suitable wetting properties and sufficient interaction with the surface of ceramic material. The shear strength achieved on metallic materials ranged from 23 to 71.5 MPa. It was the highest on copper and lowest again on Al. On the ceramic materials, it was 6 and 7 MPa as shown in **Figure 16**.

The fractured surfaces on metallic materials remained always covered with a uniform layer of solder after shear test. In case of ceramic materials, the fractured surface remained covered with a solder layer in most cases. Partial separation of solder from the ceramic substrate was observed approximately with 40% of all specimens. The fracture mostly initiated in the solder and was of ductile character. The failure took place in shear mechanism (**Figure 17**). The motion of shearing tool is clearly visible on the fracture morphology.

2.11. Active solder with lanthanum content

The research [27] was aimed at direct soldering Al_2O_3 ceramics with a copper substrate by the application of Sn2La solder. It was studied whether the Sn-based solder alloyed with La

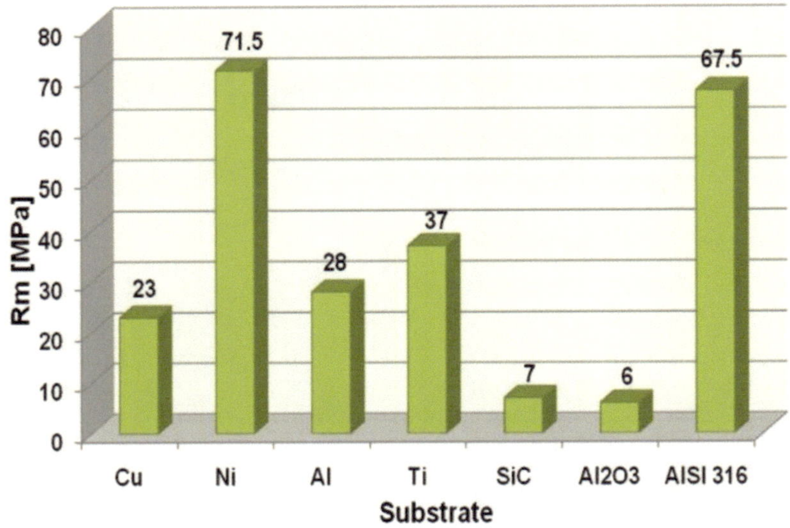

Figure 16. Shear strength of joints fabricated with solder type In70Sn30.

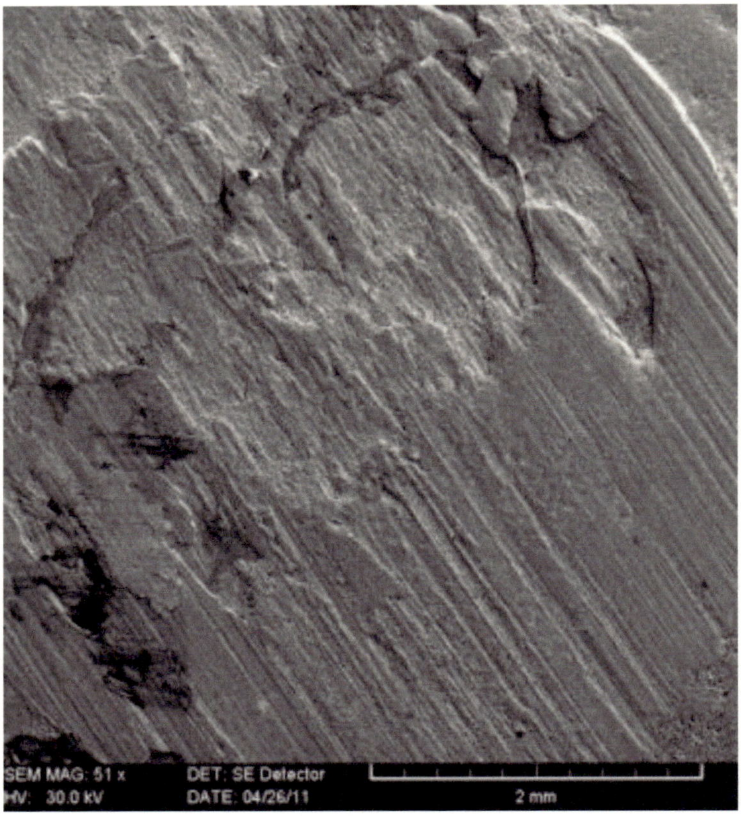

Figure 17. Fracture area of joint on the substrate of Al_2O_3.

can wet the Al_2O_3 ceramics and create thus a strong bond. The possibility to substitute La with Ti in active tin solders was also studied. For this reason, the analyses for revealing the mechanism of bond formation were performed, and the shear strength of the joints was measured. Lanthanum, as a metal with high affinity to oxygen, was applied as an active element. Soldering was performed at a low temperature in air with the application of power ultrasound. The ultrasonic equipment type Hanuz UT2 with the parameters given in **Table 3** was employed for soldering. An ultrasonic transducer was used for solder activation, which uses an oscillating piezoelectric system and a titanium tool with an outlet diameter of Ø 3 mm. The scheme of soldering with ultrasound assistance is documented in **Figure 18**. Soldering takes place through the layer of molten solder. The titanium tool, sonotrode, is thus not in a direct contact with the ceramic substrate. The soldering temperature was selected to 290°C, which is 20°C above the liquidus temperature of the solder.

Soldering procedure runs in such a manner that a layer of solder is deposited on the substrate heated at soldering temperature. The liquid solder is then subjected to active ultrasound without the application of shielding atmosphere, thus in air during the time of 5 s. After ultrasonic activation, the redundant layer of molten solder and formed oxides are removed from the substrate surface. Both substrates are prepared in the same way. The substrates with the deposited solder layer are put on each other in such a manner to create contact with the molten phase. They are then centred and the desired joint is formed by a slight compression of this assembly. Graphical representation of this procedure is shown in **Figure 19**.

Ultrasound power	**[W]**	**400**
Working frequency	[kHz]	40
Amplitude	[μm]	2
Soldering temperature	[°C]	290
Time of UT activation	[s]	5

Table 3. Parameters and conditions of soldering.

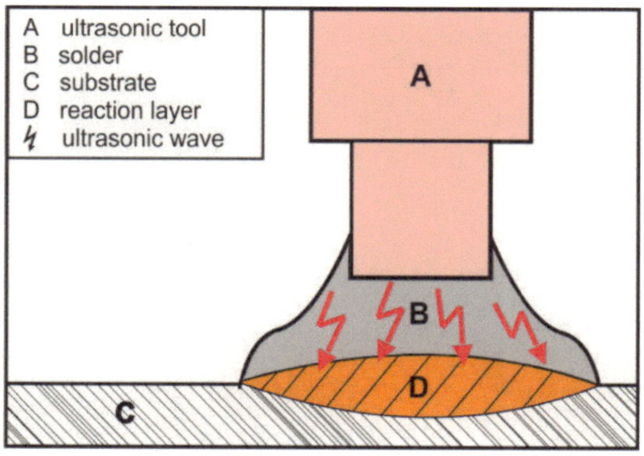

Figure 18. Ultrasonic soldering.

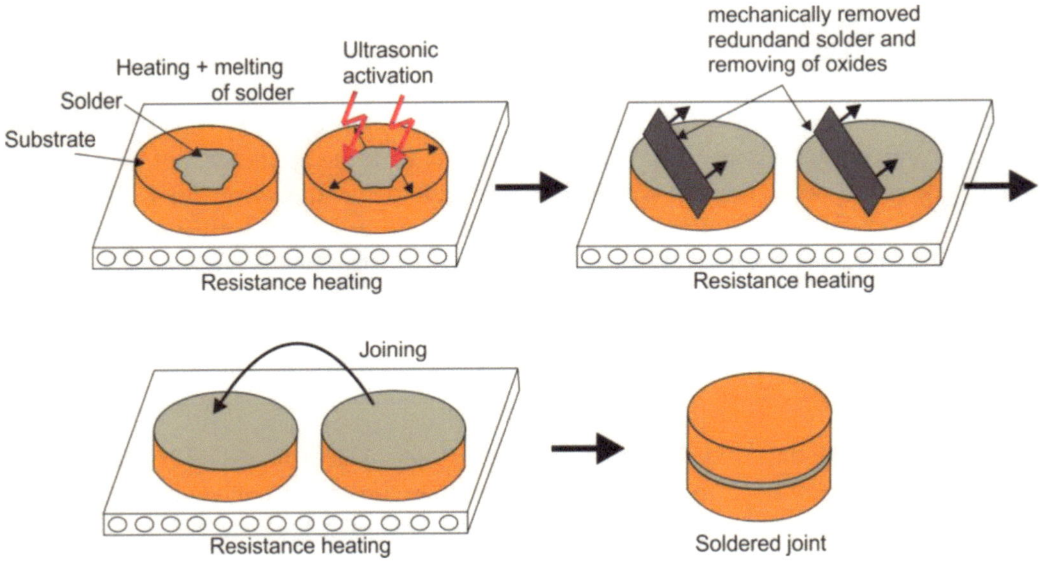

Figure 19. Procedure of joint fabrication by ultrasonic soldering.

A uniform distribution of La phases in tin matrix may be seen in the microstructure of Sn2La solder shown in **Figure 20**. No La was observed in the matrix of the studied solder. This fact was verified by the energy dispersive spectroscopy (EDS) analysis.

2.12. Analysis of interface in Al_2O_3/Sn2La solder joint

Comparison of microstructures of Al_2O_3/SiC and Cu/Cu-soldered joints from the optical microscopy is shown in **Figure 21**.

As shown, much of lanthanum is oxidized in air during ultrasonic process. The lanthanum particles are distributed to the interface with ceramic material during ultrasonic activation as shown in **Figures 22** and **23**, and they are then combined with oxides on the surface of ceramic material. The concentration line of La in **Figure 22** proves increased La concentration on the

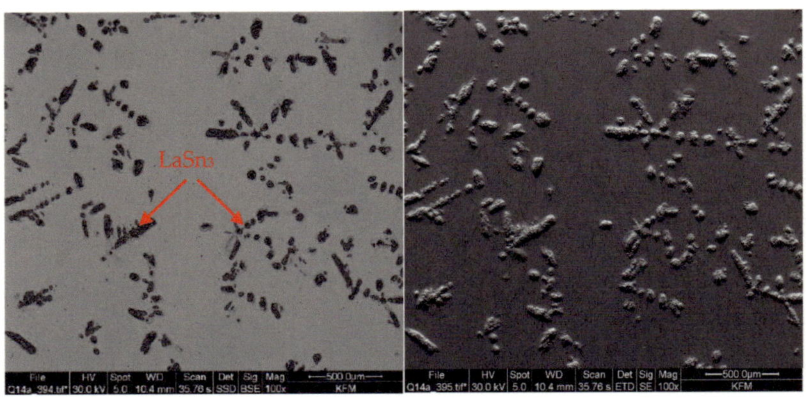

Figure 20. Microstructure of Sn2La solder.

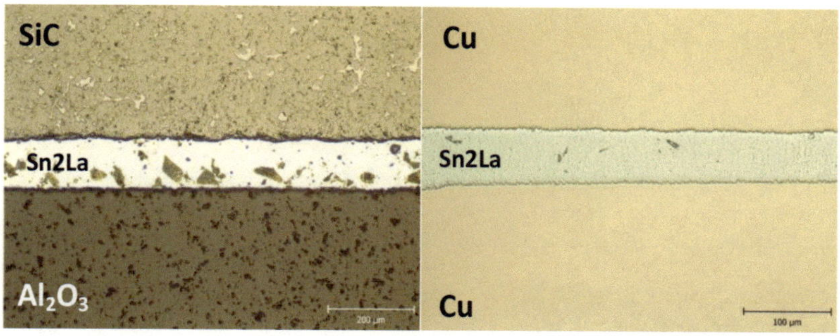

Figure 21. Comparing the microstructure of Sn2La solder after UT soldering at the same *parameters and conditions of soldering*.

interface with Al_2O_3 ceramics. A uniform, continuous layer of La oxides on the interface with ceramic material can be seen in **Figure 23**, which ensures the bond formation. The thickness of this layer is around 1.5 μm. In spite of this layer, the solder is more or less bonded to the ceramic substrate. The bond with ceramic material is of adhesion character. The formation of new intermetallic phases was not observed. This also causes lower shear strength of the bond with ceramic materials. The mechanism of bond formation is schematically outlined in **Figure 24**.

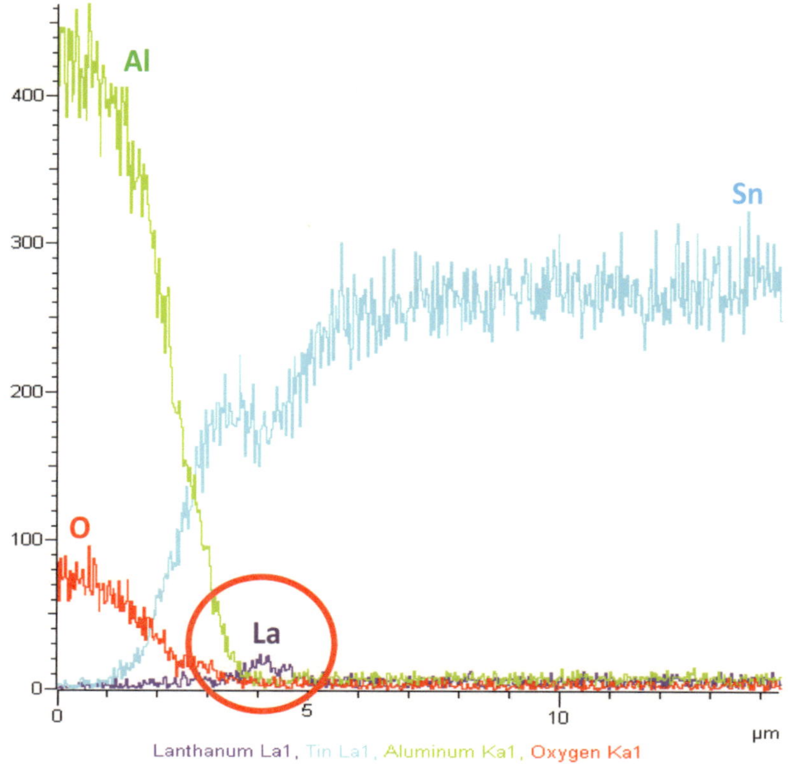

Figure 22. Concentration profiles of Al, Sn, La and O elements on the interface of Al_2O_3 ceramics/Sn2La solder.

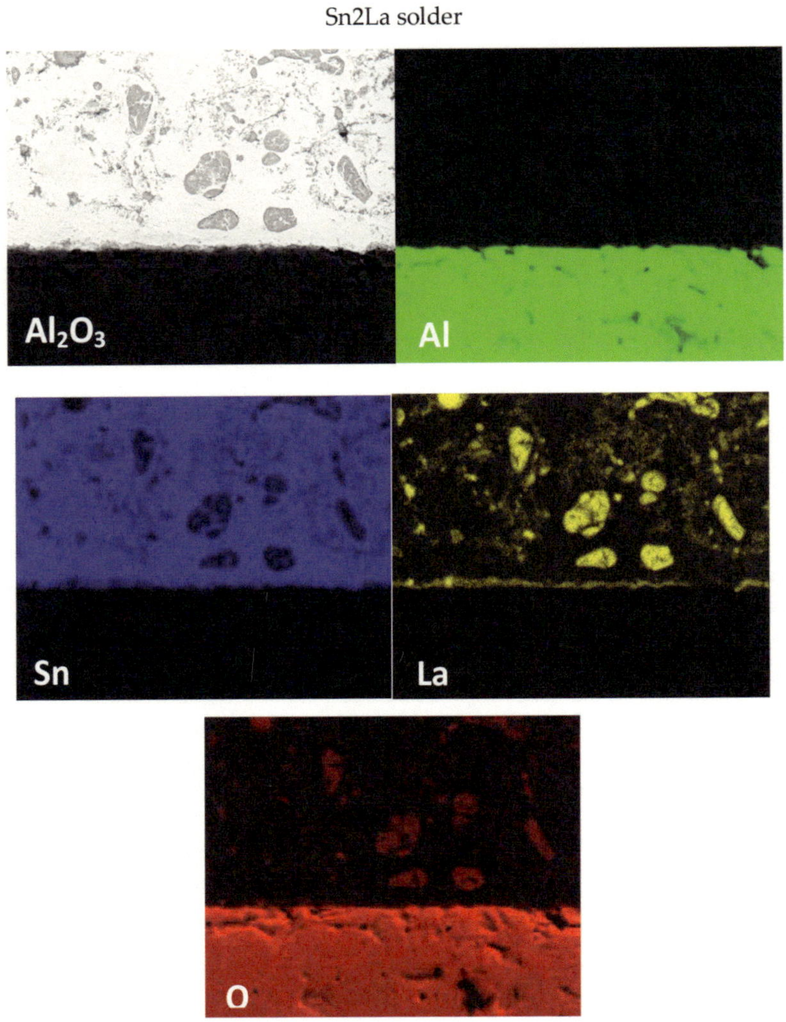

Figure 23. Planar EDX analysis of soldered interface of Sn2La/Al_2O_3.

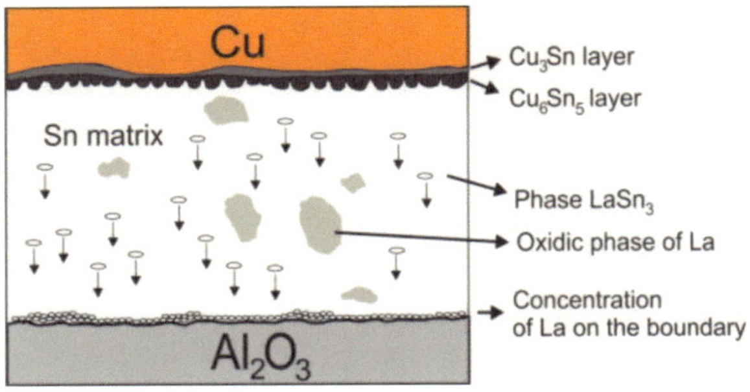

Figure 24. Mechanism of bond formation at UT activation of SnLa2 solder.

The research was primarily oriented to soldering ceramic substrate of Al_2O_3 and a copper substrate. The experiments carried out in the study of shear strength of soldered joints were extended to other metallic materials (Al, Ni, Ti and CrNi steel) and SiC ceramics in order to show broader applicability of Sn2La solder.

Measurement was performed on four specimens of each material. The results of average shear strength of joints are documented in **Figure 25**. The lowest shear strength was observed on Al_2O_3 ceramics (7.5 MPa). Little higher strength of 13.5 MPa was observed on SiC ceramics. The highest strength, when regarding the metallic materials, was achieved with Al and Ni. The strength on copper substrate was 26.0 MPa.

The Sn2La solder has shown relatively great differences in shear strength on the metallic and ceramic materials. It can be generally said that the shear strength of joints in metallic materials is almost three times higher than in the case of ceramic materials.

From the results of analysis of transition zone of soldered joints, it may be concluded that the bond with metallic material is of metallurgical-diffusion character. The bond with a ceramic material, namely Al_2O_3 (at soldering with solder containing La), is of adhesion character.

2.13. Active solder with titanium content

The aim of research [28] was to study the solderability of Al_2O_3 ceramics, silicon and copper at the application of solder type Sn-Ag-Ti activated by ultrasound. The interactions between solder, ceramic and silicon substrates were analysed. The shear strength of fabricated soldered joints was measured.

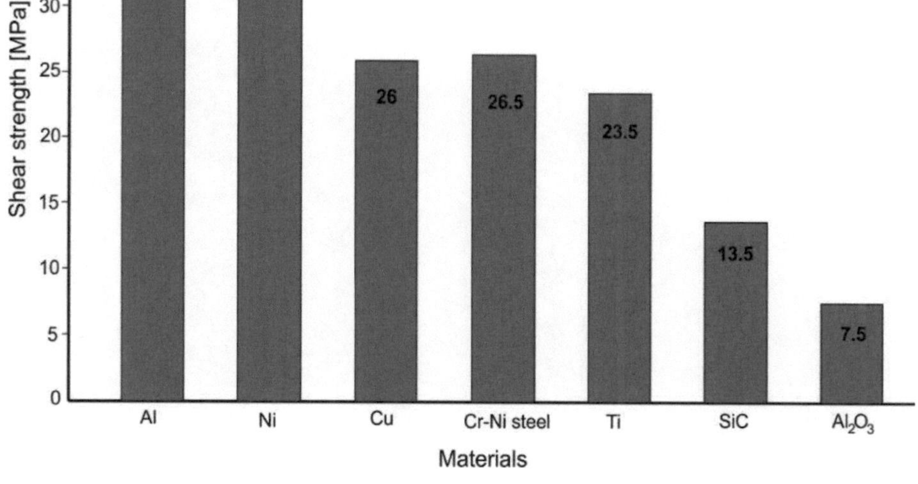

Figure 25. Results of measurements of shear strength in joints fabricated with Sn2La solder.

The Sn3.5Ag4Ti (Ce, Ga) solder was used for soldering. Soldered joints were fabricated with the application of mechanical activation by power ultrasound. Heating was realized by a hot-plate method. The soldering temperature was 280°C. The dwell time at soldering temperature was 30 s and the time of ultrasound acting was 5 s. The test specimens were prepared of Al_2O_3 ceramics, silicon as non-metallic material and Cu as metallic material.

The microstructure of solder type Sn-Ag-Ti is documented in **Figure 26**. It consists of tin matrix. The tin matrix contains unevenly distributed constituents of intermetallic Ti-Sn phases and fine needles of silver phase, Ag_3Sn, uniformly distributed along the tin grains. The presence of Ag_3Sn phase was proved by X-ray diffractometer (XRD) analysis.

XRD analysis revealed also the Ti-Sn phases. Actually identified were the Ti_6Sn_5 and Ti_2Sn_3 phases, where Ti_6Sn_5 phase was mostly represented. The formation of individual titanium phases depends on manufacturing the temperature of the solder, the amount of titanium added to solder and also the way of Ti addition to solder during its manufacture.

2.14. Analysis of soldered joint of Sn-Ag-Ti/Al_2O_3

The microstructure of Sn-Ag-Ti solder/Al_2O_3 ceramics is documented in **Figure 27**. A pronounced transition zone, reaction layer with an average thickness of 2.6 μm, is formed in the interface.

The energy-dispersive X-ray spectroscopy (EDX) analysis of chemical composition has revealed that the reaction layer (**Figure 28**) contains 5.35 wt.% Al; 37.33 wt.% Ti; 2.84 wt.% Ag and 54.48 wt.% Sn. The linear course of concentration of individual elements is documented in **Figure 26**.

During soldering process, the titanium from solder is distributed to the interface with ceramic material, where a reaction layer is formed, which ensures the wettability of Al_2O_3 ceramics. An oxidation-reduction reaction takes place between the active solder and ceramic material at the formation of reaction products, which allow the wetting of ceramics by an active solder (**Figure 28**).

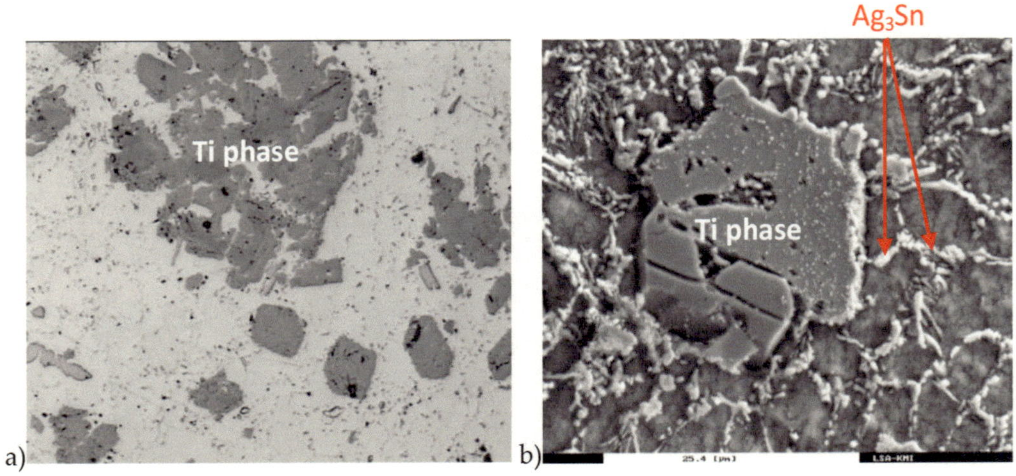

Figure 26. Microstructure of Sn-Ag-Ti solder (a) in polished condition, (b) after etching of tin matrix.

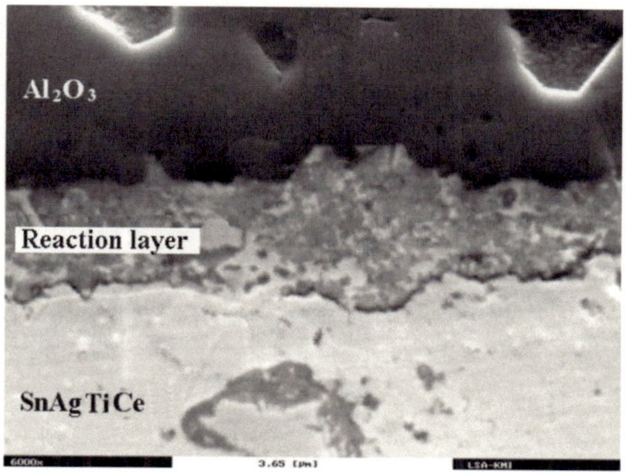

Figure 27. A detailed view of reaction layer in Al_2O_3/Sn-Ag-Ti interface.

2.15. Analysis of Sn-Ag-Ti/Si-soldered joint

The interface of Si/Sn-Ag-Ti solder joint can be seen in **Figure 29**. The Ag_3Sn phase is segregated along the grain boundaries of tin matrix of the solder. This phase increases the strength of soldering alloy type Sn-Ag-Ti. Also, the formation of a pronounced reaction layer, 1–2 μm in thickness, may be seen in **Figure 29**. The formation of this layer in the consequence of interactions is between the active element (Ti) and the surface of silicon substrate. Titanium is segregated to interface with silicon during the soldering process. The products of interaction ensure the wetting of silicon and bond formation. The diffusion mechanism takes place there. The ultrasound exerts two effects in this case, namely it speeds up the Ti diffusion and disrupts the substrate surface by the cavitation erosion.

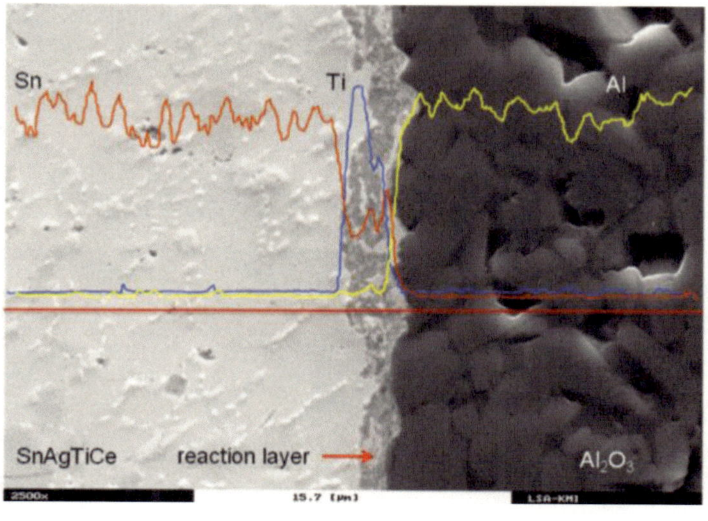

Figure 28. Interface of the soldered joint of Al_2O_3/Sn-Ag-Ti solder.

The formation and composition of reaction layer may be well observed on the planar distribution of elements attained by EDX analysis. **Figure 30** shows the map of elements in the interface of soldered joint. In **Figure 30**, we may also see the segregation of Ti on Si/Sn-Ag-Ti interface; 28 wt. % Ti was observed in the reaction zone, while the balance consisted of silicon.

2.16. Analysis of Sn-Ag-Ti/Cu-soldered joint

In case of solder type Sn-Ag-Ti, the main role in bond formation with Cu substrate is played by tin. The Cu3Sn, Cu_6Sn_5 phases were identified in the solder/Cu substrate interface in all cases, which are growing in the direction from the phase interface to solder matrix. The Cu3Sn phase is closer to Cu substrate and the Cu_6Sn_5 phase is in contact with Sn-Ag-Ti solder.

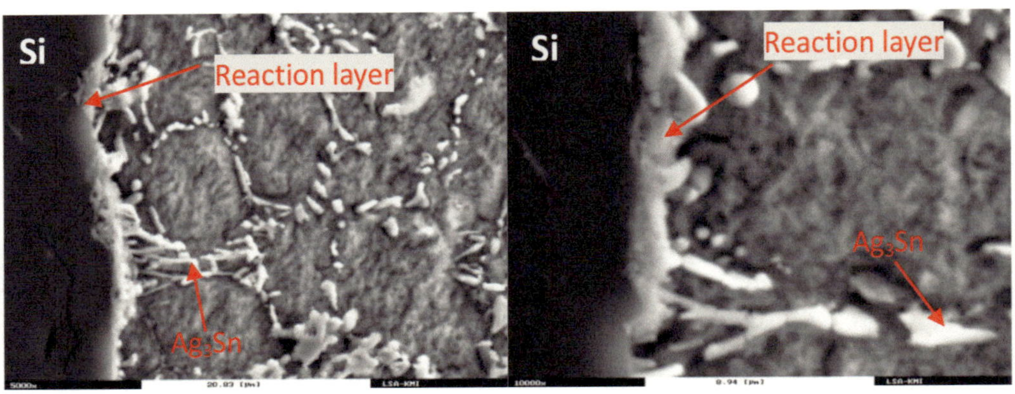

Figure 29. Reaction layer on Si/Sn-Ag-Ti solder interface.

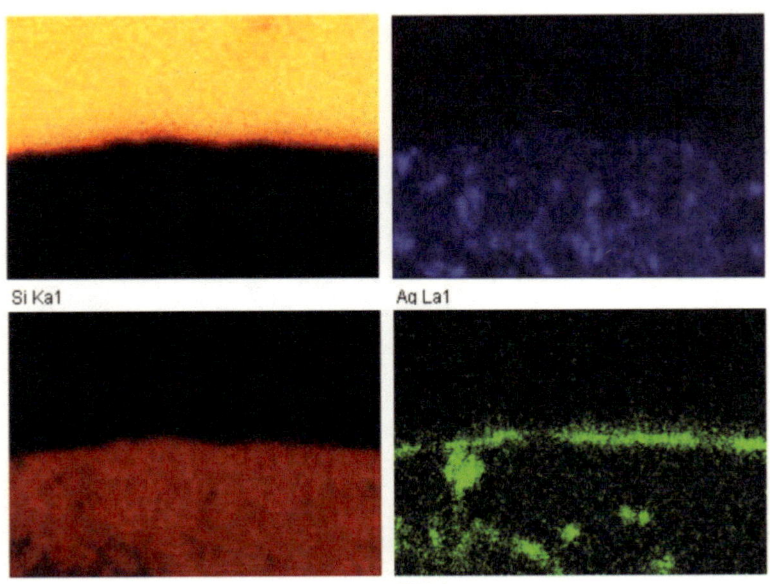

Figure 30. Map of Si, Ag, Sn and Ti elements in the interface of Si/Sn-Ag-Ti joint.

Massive transition phases of Cu_6Sn_5 were formed, up to 12 μm in width. The Cu_6Sn_5 phase has an elongated acicular shape as shown in **Figure 31**.

2.17. Shear strength of soldered joints

The research of this study was primary oriented to soldering of ceramic Al_2O_3 substrate, silicon substrate and copper substrate. However, the experiments performed in the study of shear strength of soldered joints were extended to other metallic materials (Al, Ni, Ti and CrNi steel type AISI 316) and SiC ceramics in order to approve wider applicability of Sn-Ag-Ti solder.

The measurement was performed on four specimens of each material. The results of average shear strength are documented in **Figure 32**. The lowest shear strength was observed with silicon, 23.0 MPa. Higher strength, attaining 33.0 MPa, was observed with Al_2O_3, whereas with copper the shear strength attained even 35.1 MPa.

The Sn-Ag-Ti solder exerted small differences in shear strength on metallic and ceramic materials. It may be generally said that the shear strength of joints in metallic materials is comparable with the shear strength attained on ceramic materials.

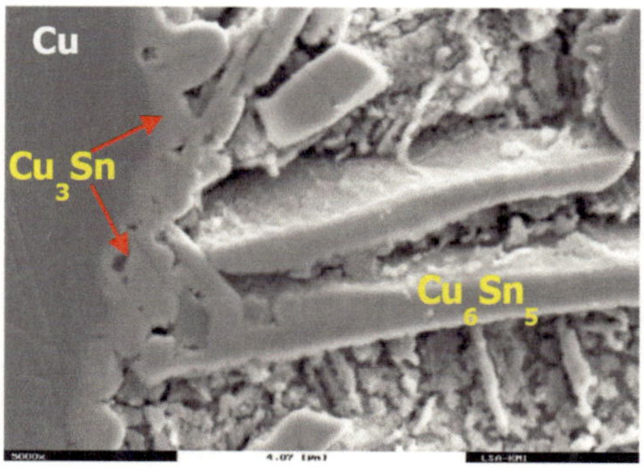

Figure 31. Analysed microstructure in interface of Cu/Sn-Ag-Ti solder joint.

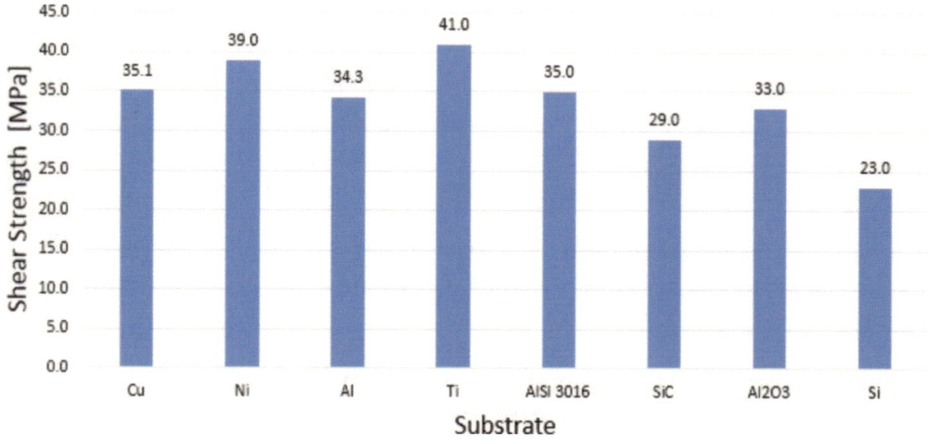

Figure 32. The shear strength of joints fabricated with Sn-Ag-Ti solder.

Acknowledgements

The contribution was prepared with the support of APVV–0023–12: Research of new soldering alloys for fluxless soldering with application of beam technologies and ultrasound, and VEGA 1/0089/17 project: Research of new alloys for direct soldering of metallic and ceramic materials.

Author details

Roman Koleňák

Address all correspondence to: roman.kolenak@stuba.sk

Slovak University of Technology in Bratislava, Faculty of Materials Science and Technology in Trnava, Trnava, Slovak Republic

References

[1] Koleňák R, Prach M. Spájkovanie/Soldering. 1st ed. Bratislava: STU Publisher; 2015. p. 285. ISBN 978-80-227-4327-3

[2] Koleňák R. Spájkovanie keramických materiálov/Soldering of ceramic materials. Zvárač/Welder. 2006;**3**(4):59-64. ISSN 1336-5045

[3] Turwitt M. Joining of Ceramics, Glass and Metal. Düsseldorf: DVS—Verlag GmbH; 1997. p. 242

[4] Ruža V, Koleňák R, Jasenák J. Spájkovanie vo vákuu/Vacuum Soldering. Trnava: SZS; 2005. p. 146. ISBN 80-969393-0-0

[5] Wlosinski W. Spájane metali z niemetalami/Bonding Metals with Non-metals. Warszawa: Panstwowe Wydawnictwo Naukove; 1989. p. 259

[6] Šváb P, Stejskal P. Způsob vytvoření kovové pajitelné vrstvy na keramických připadně skleněných substrátech/Method for Creation of Metallic Solderable Layer on Ceramic or Vitreous Substrates, Description of Invention to Authorship Certificate. ÚPV SR; 1981

[7] Koleňák R. Fyzikálno-metalurgické aspekty spájkovania keramických materiálov s kovmi/Physico-metallurgical aspects of soldering ceramic materials with metals [thesis]. Trnava: KZv MtF STU; 2001. p. 145

[8] Koleňák R, Turňa M, Vaščák M. Vytváranie tenkých povlakov PVD naprašovaním/Creation of thin coatings by PVD sputtering. In: CO-MAT-TECH 2000, STU Bratislava; 2000. pp. 145-151. ISBN 80-227-1413-5

[9] Iždinský K, Ivan J, Ruža V. Charakteristika procesu vzniku spojenia pri spájkovaní kovov s keramikou pomocou aktívnej spájky Ag-Cu-Ti/Characteristic of bond formation process in soldering metals with ceramics by use of an active solder type Ag-Cu-Ti. In: Zborník z konferencie Nové materiály a technologie/Proceedings from the Conference New Materials and Technologies, Ostrava - Trojanice; 1995. pp. 126-129

[10] Koleňák R. Solderability of metal and ceramic materials by active solders. In: Forschungszentrum. 1st ed. Dresden; 2008. p. 72. ISBN 978-3-941405-03-5

[11] Dupák J, et al. Pájení tvárnými aktivními pájkami/Soldering with ductile active solders. Zváranie/Welding. 1999;**48**(6):130-132

[12] Schwartz MM. Handbook of Structural Ceramics. New York: McGraw-Hill; 1992. p. 949. ISBN 0-07-055719-5

[13] Hillen F, Rass JI, Lugscheider E. Low temperature joining for heat exchanger and other applications. In: ISBC 2000, ASM International, Albuquerque, New Mexico; 2000

[14] S-bondTM Handbook, STB Copyright, USA; 2002

[15] Pulc V, et al. Možnosti využitia Ag-Cu-Ti aktívnej spájky na spájanie keramiky na báze Si_3N_4 s titánom/Possibility of application of an active Ag-Cu-Ti solder for bonding Si_3N_4-based ceramics with titanium. Zváranie/Welding. 1994;**43**(2):30-34

[16] Ruža V. Pájení/Brazing-Soldering. 2nd ed. Prague: SNTL; 1988. p. 456

[17] Schwartz M. Brazing. London: Chapman and Hall; 1995. p. 399

[18] Ljungberk L. Joining of Ceramics to Metals by Brazing. Göteborg: Chalmers University Library Reproservice; 1992. p. 112

[19] Koleňák R, Ruža V. Spájkovanie materiálov/Brazing-Soldering of Materials. Bratislava: STU; 2007. p. 151. ISBN 978-80-227-2705-1

[20] Humpston G, Jacobson MD. Principles of Soldering and BrazingASM International; 1993. p. 281

[21] ASM Handbook, Vol. 6: Welding, Brazing and Soldering. OH; 1995

[22] Indium Corporation of America. Bonding Non-Metallic Materials Using Indium and High Indium Alloys (Application note). 2016. Available from: www.indium.com

[23] Provazník M, Koleňák R. A study of the interface of soldered joints of SnlnAgTi active solder with ITO ceramics. Acta Polytechnica. 2010;**50**(6):70-73. ISSN 1210-2709

[24] Koleňák R, Martinkovič M. Shear strength of joints fabricated by solders with high indium content. In: Materials Science Forum: Metallography' 2013: 15th International Symposium on Metallography. Slovak Republic, Stará Lesná, 24-26 April 2013. pp. 461-464. ISSN 0255-5476

[25] Chuang RW, Lee CC. High-temperature non-eutectic indium-tin joints fabricated by a fluxless process. Thin Solid Films. 2002;**414**:175-179

[26] Zhang X, Xu G, Guoi F. Effects of micron-sized metal particles on the mechanical properties of In-Sn thermal interface materials. In: International Conference on Electronic Packaging Technology & High Density Packaging, 2011. pp. 1169-1171

[27] Koleňák R, Kostolný I. Study of direct bonding ceramics with metal using Sn_2La solder. Advances in Materials Science and Engineering. 2015;**2015**:Article ID 269167, 13 p. ISSN 1687-8434

[28] Koleňák R. Solderability of ceramic and metal materials with utilization of active solders and power ultrasound. International Review of Mechanical Engineering. 2016;**10**(6):405-408. ISSN 1970-8734

Chapter 5

Evolution of Pb-Free Solders

Wayne Ng Chee Weng

Additional information is available at the end of the chapter

http://dx.doi.org/10.5772/intechopen.69553

Abstract

This chapter discusses the evolution of lead free (Pb-free) solder, from tin-silver-copper (SAC) system with silver content of 3.0, 3.8, and 4.0 to low SAC system such as SAC0307 and SAC105 and the emerge of high reliability Pb-free solder. The discussion covers the reason and the driving force of industries implementing this change. The solder composition has evolved further recently to fulfill high reliability requirement of certain sectors such as automotive, aerospace, and military which are preparing to go green in soldering technology. This kind of high reliability solder involves additional microalloying of tin (Sn)-based solder in making it to be more robust. In this chapter, the author will introduce the techniques used by solder makers and researchers in enhancing the Pb-free solder strength in the recent evolution. Recently, attention has been drawn to low temperature joining technology again such as silver sintered joint and liquid-phase diffusion bonding material used in high power density and high junction temperature-integrated circuits. Pb-free joining material is required to replace the high Pb solder, which is still commonly used in such high-power devices.

Keywords: Pb-free solders, particle strengthening, Ag3Sn coarsening, solid solution strengthening, high temperature die attach material, silver (Ag) sintering, liquid-phase diffusion bonding (LPDB)

1. Introduction

The adoption of Restriction of Hazardous Substances Directive 2002/95/EC (RoHS 1) in 2003 was one of the driving forces mandating the electronic industries to look for alternatives to replace the tin-lead eutectic alloy (SnPb-6337), which had been used for decades. RoHS 1 has limited the content of Pb to be less than 1000 ppm by weight in each homogenous material, which is including the soldering material. The SnPb-6337 contains 37% of Pb, which is an element with known harmful effect to human health and the environment. Thus, SnPb-6337 has to be replaced in order to meet the RoHS compliance. Pb is just one of the six banned substances under RoHS 1 [1]. The banned list of RoHS has expanded to 10 under Directive (EU) 2015/863 (RoHS 2), which was published in 2015. The full list of banned substances in RoHS 2 is shown in **Table 1**. PBB and PBDE are mainly used as frame retardant in plastics and hexavalent chromium is used in chrome plating, chromate coatings and primers, and in chromic acid. DEHP, BBP, DBP, and DIBP are phthalates, which are being used as plasticizers. These substances are rare in solder materials. Out of the 10 substances listed in RoHS 2, only Pb is the most concerned substance for solder industry. This explains why some of us refer RoHS compliance solder as Pb-free solder. Although the RoHS compliance cannot help in reducing the electronics waste, which has increased tremendously in recent years, it helps to minimize the hazardousness of such trash to environment.

Since 2006, Pb-free soldering has been in the main stream of industries because RoHS 1 took effect on 1st July 2006. Many changes in alloy composition are adopted in order to match the requirement of solder joint with Pb-free definition. Tin Copper (SnCu), Tin Silver (SnAg), and Tin Silver Copper (SnAgCu or SAC) are the dominating Pb-free solder alloys adopted by the industries.

SAC alloy had received most publicity in the initial stage of conversion from SnPb-6337 to Pb-free solder. One of the reasons was because SAC alloy with 3.0% silver (Ag) and 0.5% copper (Cu), SAC305 was endorsed by the IPC Solder Value Product Council as the preferred option for SMT assembly. Other than SAC305, the industries had also adopted other SAC alloys with higher silver content such as Sn3.8Ag0.7Cu (SAC387) and Sn4.0Ag0.5Cu (SAC405), which are believed to be the true eutectic ternary solder alloys, which have single melting point at 217°C. SAC305 is a hypo-eutectic ternary alloy with melting range of 217–219°C. However, the high cost of Ag compels the industries to consider other alternatives such as low silver solution, SAC0307, or no silver solutions such as SnCuNi (SN100C®). A more detailed discussion on this topic will be covered in Section 2 of this chapter.

Substance	RoHS limit
Lead (Pb)	1000 ppm/0.1%
Mercury (Hg)	1000 ppm/0.1%
Cadmium (Cd)	100 ppm/0.01%
Hexavalent chromium (Cr^{6+})	1000 ppm/0.1%
Polybrominated biphenyls (PBB)	1000 ppm/0.1%
Polybrominated diphenyl ether (PBDE)	1000 ppm/0.1%
Bis(2-ethylhexyl) phthalate (DEHP)	1000 ppm/0.1%
Butyl benzyl phthalate (BBP)	1000 ppm/0.1%
Dibutyl phthalate (DBP)	1000 ppm/0.1%
Diisobutyl phthalate (DIBP)	1000 ppm/0.1%

Table 1. Ten substances which are banned in Restriction of Hazardous Substances Directive 2015/863 (RoHS 2).

The requirement of solder joint performance is getting stringent in order to support the modern electronification. It is especially for automotive industry. In a modern car system, the usage of advanced electronic components is getting common, for example, in logic control, switches, and sensors. The adoption of electronic components has significantly boosted the speed and accuracy of these systems. The electronification of car system such as advanced driver assistance systems (ADAS) has given another level of driving experience to consumer. On the other hand, the incorporation of multimedia system in car system, which is to increase the comfort level of driver has further increased the usage of electronic components in a car. It explains the growth of auto electronics has surpassed other industries in recent years [2]. Due to the harsh use environment where the under-hood temperature is much higher than other applications, the requirement on the solder joint performance is higher. Moreover, these systems are considered as life-critical system, which has increased the challenges in selecting an appropriate solder alloy. Due to this kind of new requirements, solder manufacturers and researchers are striking to invent new alloy, which can support this platform. It also implies that the conventional Pb-free solder alloys such as SnCu, SAC, and low SAC are showing limitations in fulfilling the demand of such applications. The various strengthening mechanisms in Sn-based solder alloy which have been deployed by the industries will be discussed in Section 3 of this chapter.

Besides in-car system, the effort of popularization of electric vehicles has further promoted the usage of electronic components in automotive industry. In view, the fact that this kind of system involves high current and high temperature, conventional SAC Pb-free alloys cannot fulfill the basic requirement. This is the reason why the exemption for RoHS compliance has been further extended to 2021 for Pb use in high-melting temperature type solders (i.e., Pb-based solder alloys containing 85% by weight or more Pb, Category 7a). Nowadays, most of the electronic devices for such application are still adopting high Pb solder as the die attach material. There are various options available in replacing high Pb solder at die attach process. But, there are still gaps to be filled up before such new materials can be fully commercialized and mass adoption of such new materials in current production.

Internet of things (IoT) has caught high attention from industries. Many industry players believe that the IoT will boost the growth of electronic component to another level, the era of IoT. The IoT has reshuffled the business process of many big organizations in preparing themselves to ride on this wave. To interpret the IoT in a simpler way, it is an application which allows the wide use of sensor in our daily life. In IoT, there is also a platform, which can connect all the relevant sensors for data analyzing and actuation. The objective of IoT is to realize a smarter world. According to *Business Insider*, there are 1.9 billion interconnected devices in 2014, with an expected 9 billion by 2018. By then, the number of IoT devices will surpass the summation of mobile phone, tablet, and PCs, which are the current major consumer market. With such high quantity of IoT device, the usage of solder as interconnect is also prospective. The selection of solder alloy to fulfill the requirement of this sector will focus in meeting the specific criteria for manufacturing, reliability, toxicity, cost, and availability.

In view of prospective growth of electronic component and new solder joint requirement, solder manufacturers and researchers are working together in developing new alloy or modifying existing alloys to close the gap between the user and supplier of solder material. In the chapters that follow the evolution of Pb-free alloy and the techniques used in enhancing solder alloy strength are discussed.

2. Evolution from high silver to low silver solder

In 2000s, SAC system had been identified as replacement for Sn-Pb eutectic solders in fulfilling the RoHS compliance. Different countries or organizations had chosen different SAC composition as their preferred solder composition. They are Sn3.0Ag0.5Cu, SAC305 (Japan), Sn3.5Ag0.9Cu, SAC3509 (European Union), and Sn3.9Ag0.6Cu, SAC3906 (USA) [3].

In 2000, it was the era of consumer product, which had incorporated many electronic components in its system such as personal computer, television, radio, rice cooker, and window-mounted air conditioner. These products were mostly using through-hole components and/or big surface mount technology (SMT) components. It implies that the consumption of solder per unit device was huge during that period of time. It explains the reason of mass adoption of SAC305 instead of other Pb-free solutions. It is because the lower silver content of SAC305 as compared to other SAC alloys mentioned in former paragraph. Silver is a precious metal and its inclusion in solder alloy does increase the cost of solder significantly. For example, to produce 1 kg of SAC305, 1 ounce of silver needs to be added. It depends on the market price of silver, assuming USD20 per ounce, the cost of silver in 1 kg of SAC305 is almost 50% of the selling price of 1 kg of SAC305 solder bar. When the silver price surges, for example, in April 2011, when the silver price was USD 40 per ounce, the cost of silver of a SAC305 solder bar was more than 50% of its selling price. For those compositions with higher silver content, the ratio of silver cost against solder bar selling price is usually more than 50% depending on the tin price. It becomes a big burden to solder users. Solder bar is mainly used in wave soldering process. It was a very common process in printed circuit board assembly (PCBA) for those consumer products mentioned above. During the peak of this era, the consumption of solder bar could exceed 2000 kg per month for a single customer who has more than 10 wave soldering lines in production. Another reason of mass adoption of SAC305 instead of other higher Ag bearing SAC alloy is more a political reason. During that time, Japanese consumer product was very popular and common. *National* brand electric rice cooker should be one of the must have electrical appliances in every home. SAC305 was the Japan Electronics and Information Technology Industries Association, JEITA recommended Pb-free alloy too.

Nevertheless, the industries were looking for alternatives for SAC305 very quickly after its introduction to industries, even though with its acceptance as Pb-free alloy in replacing SnPb-6337 solder. In general, there are two major reasons in justifying this direction, cost and drop impact resistance of SAC305 alloy. The users were looking for alternative to mitigate the high cost of SAC305 as mentioned above. This was the time when low SAC alloy started to join the supply chain of solder. **Table 2** lists some of the low SAC alloys available in market offered by different solder suppliers. The introduction of these low SAC alloys was as soon as 2 years after the mass adoption of SAC305 alloy as Pb-free solder in the industries. Nevertheless,

SAC305 is still the major adopted alloy in the industries after 10 years of RoHS enactment. It is worth to note that Nihon Superior has introduced Ag-free Pb-free solution since 1999. It has become a main solution in soldering especially in wave soldering and Pb-free hot air solder leveling (HASL) process.

With lower silver content in SAC alloy, some users were hoping the hot tear or shrinkage cavity could be mitigated. This phenomenon which was first believed as a process defect was found when users migrating from SnPb-6337 to SAC305. To date, many users are still confused by this phenomenon even though IPC has classified it as an acceptable phenomenon. **Figure 1** shows the comparison of few low SAC alloys and a eutectic Tin Copper Nickel alloy.

Besides the appearance of solder whose difference is easily noticeable, the users were also concerned about the reliability of low SAC especially the intermetallic compound (IMC) formed in between the solder and soldering pad. **Figures 2** and **3** show the IMC of SAC 305, low SAC, SAC1205, and Ag-free Pb-free solder (SnCuNi) on Copper (Cu) pad and Nickel Gold (NiAu) pad, respectively. Both SAC alloys regardless of Ag content were showing columnar growth of IMC and the IMC continued to grow during the heat treatment at 125°C for 500 h. The SnCuNi IMC appeared to be flat and stable even after the heat treatment. Besides the columnar IMC, crack was observed at the Cu pad after heat treatment within the IMC of SAC305 but not at the SAC1205 and SnCuNi alloys. This observation matched the explanation proposed by Nogita et al. in his journal [4]. The Ni addition in Pb-free alloy has successfully suppressed the polymorphic transformation of Cu_6Sn_5, which is the culprit of this IMC crack. There is transformation of hexagonal to monoclinic Cu_6Sn_5 and vice versa at temperature of 186°C due to the allotropic attribute of Cu_6Sn_5. This transformation involves volume change of the Cu_6Sn_5 IMC, which causes high stress at this layer. The stress has exceeded the strength of this IMC and caused the crack within this layer. Micro crack might have formed within the SAC305 IMC during the solidification process of joint formation and the cracks propagated during the heat treatment due to IMC growth. This has explained why the crack still formed even though the heat treatment temperature was just 125°C, which was far below the polymorphic transformation temperature.

Solder manufacturer	Alloy composition	Solidus (°C)	Liquidus (°C)	Debut (year)
Alpha Assembly Solutions	Sn0.8Ag0.7Cu	216	225	2007–2010
	Sn0.3Ag0.7Cu	217	228	2007–2010
Senju Metal Industry	Sn0.3Ag0.7Cu	217	227	2007–2010
	Sn1.0Ag0.7Cu	217	224	2007–2010
Indium Corporation	Sn1.0Ag0.5Cu	217	225	2007–2010
	Sn0.3Ag0.7Cu	217	227	2007–2010
Nippon Micrometal Corporation	Sn1.2Ag0.5Cu0.05Ni	217	227	2000s
Nihon Superior	Sn0.7Cu0.05Ni0.01Ge	227	227	1999

Table 2. List of low SAC solder and no Ag solder as alternatives of SAC305.

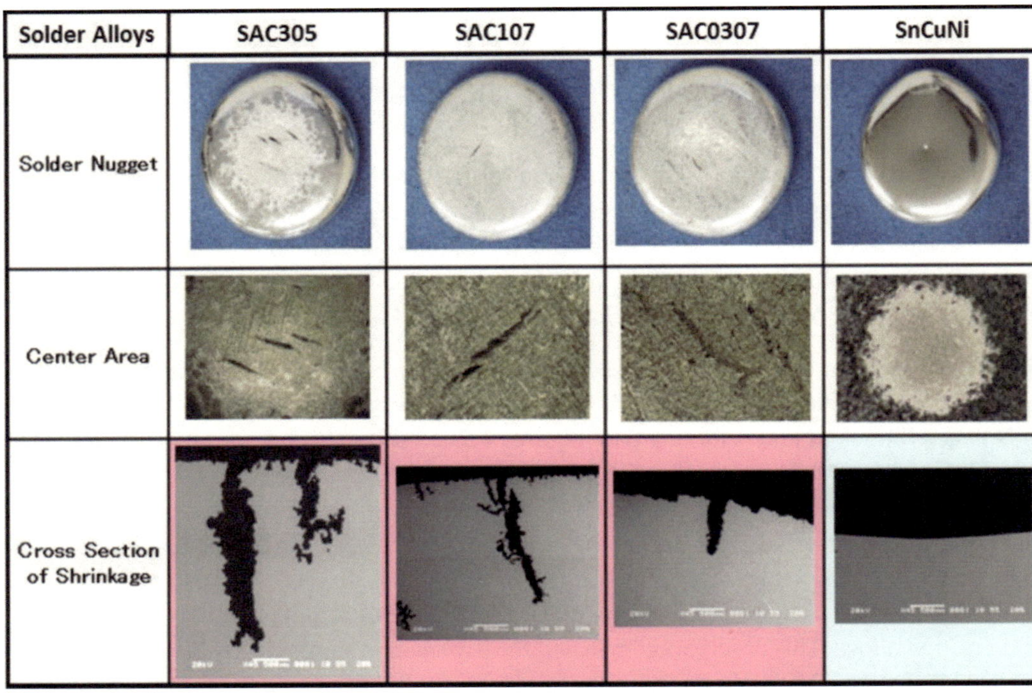

Figure 1. Appearance comparison of different SAC alloys and a eutectic SnCuNi alloy.

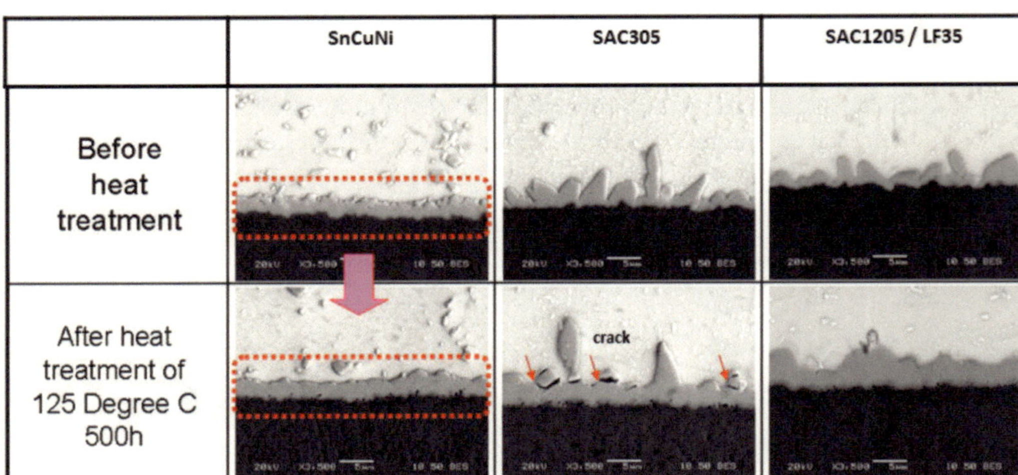

Figure 2. IMC formed on Cu pad with solder alloy of SnCuNi (SN100C®), SAC305 and SAC1205 (LF35).

Besides the cost concern mentioned earlier, another driving force which triggered the industries to consider low SAC alloy was the drop impact resistance. It is widely accepted that silver addition can effectively reduce the liquidus and increase the yield strength and modulus of SAC alloy; also because of the high strength and high modulus which make the alloy

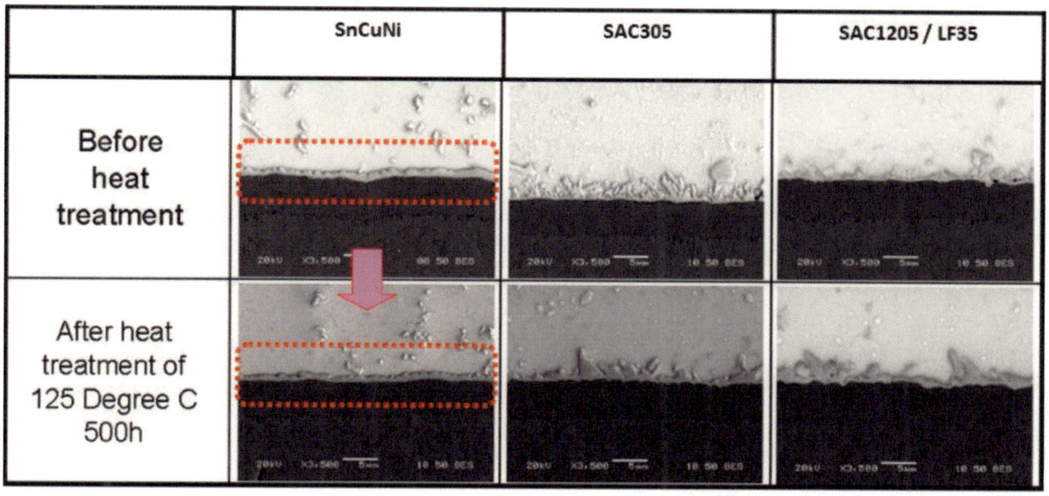

Figure 3. IMC formed on NiAu pad with solder alloy of SnCuNi (SN100C®), SAC305, and SAC1205 (LF35).

readily transfer the stress to the solder joining interface which is the region of IMC [5]. Due to this mechanism, most of the failures found in drop impact resistance test with conventional SAC alloys mostly locate at the IMC region, and most of these failures are early failures. By reducing the Ag content of SAC alloy, the modulus of alloy decreases. It has significantly increased the resistance to the drop impact. The urge of looking for an alloy with high drop impact resistance increased further due to the popularity of portable devices in mid-2000 to late-2000. During that period of time, portable music player such as MP3 player and iPod and handheld game console were very common. The robustness against drop shock was crucial to keep a reasonable reliability of such devices. For example, during the initial implementation of Pb-free soldering, most of the solder ball used in ball grid array, BGA package is SAC305 composition. Many mobile integrated circuits, ICs had migrated to LF35 or SAC1205 BGA solder ball in order to increase the drop test performance of the ICs. However, by only lowering the Ag content of SAC alloy did not give a total solution to meet the expectation of users. Many other microalloying had been deployed by solder manufacturers in order to increase the alloy performance of low SAC alloy such as adding Nickel, Ni, Zinc, Zn and Manganese, Mn into the alloy system. As shown in **Figure 2**, the addition of Ni into LF35 composition has suppressed the polymorphic transformation of Cu_6Sn_5 during the solidification. It cannot be achieved by only reducing the Ag content of SAC alloy.

The acceptance of low SAC or SnCuNi as Pb-free solution is also encouraged by the readiness of other material to Pb-free soldering process. One of the major challenges in adopting low SAC or SnCuNi alloy is the high melting point of this alloy. It has approximately 44 and 10°C higher melting point comparing to eutectic SnPb alloy and SAC alloy, respectively. However, the market is now more ready for this Pb-free soldering. The reflow and wave soldering machines are having higher heat capacity to accommodate the high melting point of Pb-free solder, PCB, and electronic components are using higher glass transition (Tg) temperature materials and soldering flux can survive longer in high-temperature reflow. All these improvements have enabled the adoption of low SAC and SnCuNi alloy in soldering. This has also explained the market share of this low SAC and SnCuNi alloys is expanding in recent years. SN100C® has become one of the preferred choices in Pb-free wave soldering process.

In short, there are two driving forces, which have triggered the proliferation of low SAC and SnCuNi alloys for electronic industries. Someone has named the low SAC as second generation of Pb-free alloy. **Figure 4** summarized the evolution of this Pb-free alloy from conventional SAC to low SAC alloy for electronic industries. This transition is very important to support the IoT era. One of the characteristics of IoT is huge in quantity. The high usage of electronic components in IoT era also prompts the usage of solder. The type of solder for this era must be able to fulfill the specific requirement in manufacturability, reliability, toxicity, cost, and availability. The second generation Pb-free alloy should be a better choice for the IoT era.

The following section discusses the development of Pb-free alloy into third generation alloy for high reliability application.

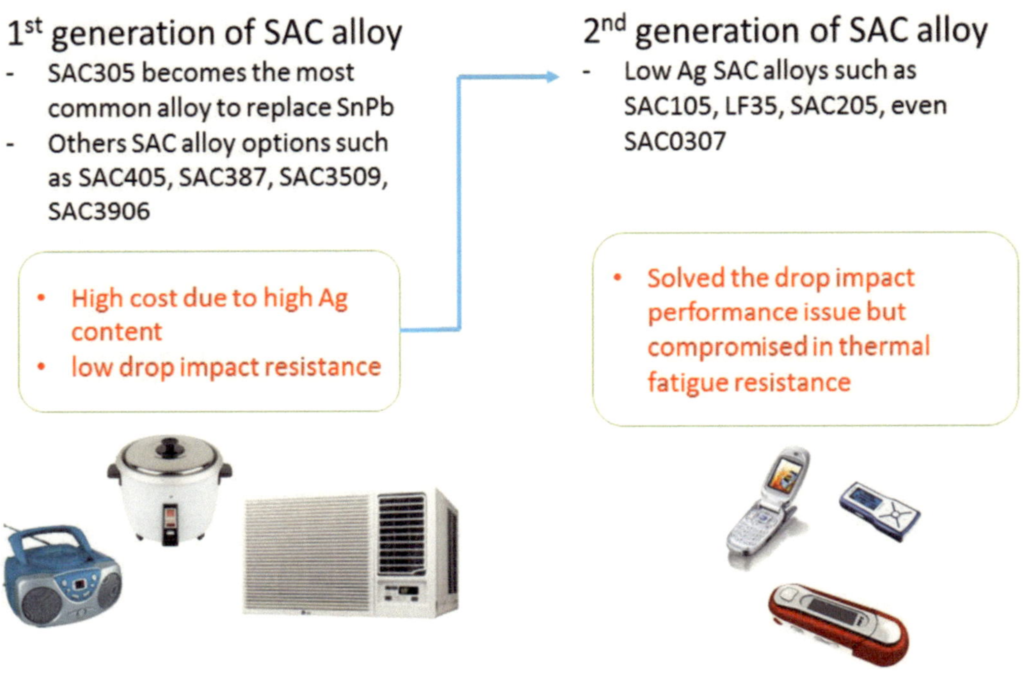

Figure 4. Evolution of SAC alloys in Pb-free solution from first generation to second generation.

3. High reliability alloy is required

As discussed in Section 2, there is an evolution of Pb-free solder from conventional high Ag bearing SAC alloy to low SAC alloys because of cost and drop impact resistance. This is

http://dx.doi.org/10.5772/intechopen.69553

the transition of first-generation SAC alloys to second-generation SAC alloys. However, the lower Ag content has compromised in thermal fatigue performance. It is because the strength enhancement of SAC alloy is through the dispersion of very small Ag_3Sn IMC within the tin matrix. This fine Ag_3Sn IMC has the effect in inhibiting the dislocation movement of the tin matrix. This mechanism makes the SAC alloy to have higher modulus as compared to the SnPb-6337. The more Ag content will increase the amount of fine Ag_3Sn IMC, but it also increases the possibility of large size Ag_3Sn platelet growth. This large Ag_3Sn platelet can be the stress concentration point and crack will initiate from this point. In extreme case, the large Ag_3Sn plate can even protrude out the solder surface and deform the small solder joint [5]. Therefore, the large Ag_3Sn platelet is not a desirable morphology in a solder joint. However, the lower Ag content in low SAC alloys has tremendous reduction in amount of Ag_3Sn IMC. Although there is another IMC, which is the Cu_6Sn_5 within the bulk solder, this IMC is much fewer compared to Ag_3Sn IMC of conventional SAC alloys. Moreover, a portion of the Cu of bulk solder will be extracted in forming the IMC at Cu pad. This further reduces the amount of Cu_6Sn_5 within the bulk solder. Thus, the main strength of the SAC alloy is still the dispersion of fine Ag_3Sn within the tin matrix.

The compromise in thermal fatigue performance of SAC alloy has been further complicated by the phenomenon of Ag_3Sn IMC coarsening. The Ag_3Sn coarsening is a process of Ostwald ripening. This is an evolutional process of an inhomogeneous structure where the small particle is growing and drawing the adjacent particle to form a bigger particle over time. This is a spontaneous process because big particle is more energetically stable than small particle. It is because the internal pressure is reversely proportional to the radius of a particle. Hence, the small particles have higher surface energy. The small particles will try to achieve a higher energetic stability by growing to bigger size. This is coarsening. Besides that, the Ostwald ripening is a thermodynamically driven process. If the environment temperature is high, the process of coarsening will increase. This Ag_3Sn coarsening has significant impact to the strength of SAC alloy. As discussed earlier, the strength of SAC alloy is the dispersion of fine Ag_3Sn IMC. **Figure 5** shows an example of the Ag_3Sn coarsening within SAC305 alloy after exposed to high temperature, 125°C for 6 months.

In the coarsening process, two changes will be observed: 1. The amount of Ag_3Sn in the tin matrix, and 2. The distance between Ag_3Sn IMCs. When the amount of Ag_3Sn shrinks and the distance between Ag_3Sn gets wider, its ability in inhibiting the dislocation movement reduces. This is a degradation process of SAC alloys. In fact, we can easily demonstrate this mechanism by exposing SAC alloy to high temperature environment over a period of time. The tensile strength of SAC alloy reduces after the thermal aging. **Figure 6** shows an example of this degradation in SAC305 alloy. This is an experimental data of the comparison of tensile strength of tensile test specimen made of SAC305 alloy with and without exposure to isothermal aging (150°C for 500 h).

3.1. High reliability solder alloy for PCBA

The degradation of SAC alloys over time is worrying because it is a challenge for reliability engineer to predict the life of a solder joint. It is especially critical on those life-critical system

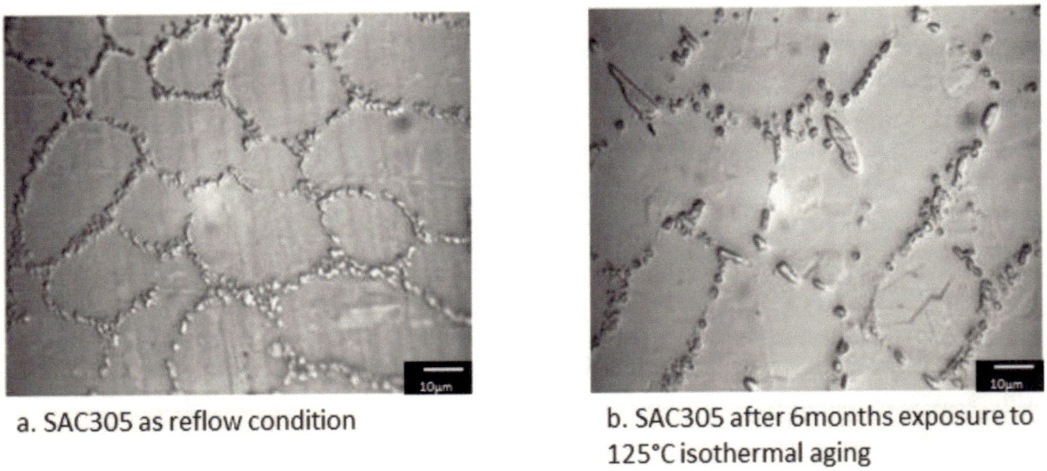

Figure 5. Ag_3Sn coarsening in high temperature storage test.

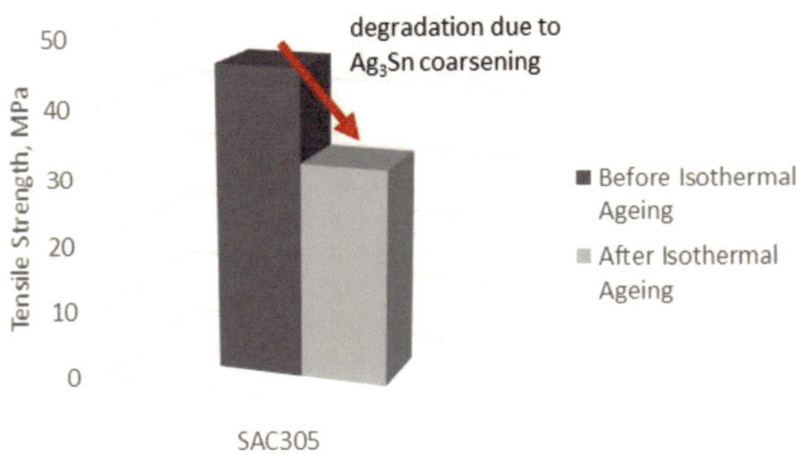

Figure 6. Tensile strength of SAC305 before and after isothermal aging at 150°C for 500 h.

such as the electronic system in a car or air plane. As mentioned in Section 1, the growth of auto electronics industry is very aggressive recently. This industry demands high reliability and zero defect electronic components. There are some original equipment manufacturer (OEM) parts that have long guaranteed product life ranging from 10 to 15 years. This is the reason why in some of the automotive applications, the SnPb-6337 is still being used. Since Pb-free soldering is getting pervasive even in these kind of high-end applications and SAC305 is still the most common Pb-free alloy, which apparently cannot fulfill the high reliability requirement of such applications, great driving force arises to drive solder manufacturer and researchers to identify new solution of Pb-free alloy.

Besides the Ag_3Sn coarsening, the homologous temperature of solder system becomes greater due to higher application environment temperature. In order to keep the strength especially in fatigue, solder with lower homologous temperature is needed. The under-hood electronic components are a good example of this case. It has further justified the need of new alloy to support future electronics reliability needs.

There are consortiums and nonprofit organizations such as Universal's Advanced Research in Electronics Assembly (AREA) Consortium and International Electronics Manufacturing Initiative (iNEMI) gathering the experts of solder to jointly identify new Pb-free alloy in solving the issues mentioned above [6]. This new alloys are named as third generation of Pb-free alloy, which has superior performance in both thermal fatigue resistance and drop impact resistance to fulfill the high demand of solder joint reliability. **Figure 7** summarizes the evolution from first generation to third generation of Pb-free alloys.

Apparently, more strengthening of Pb-free alloys is required in order to improve the strength. As discussed earlier, SAC alloys are using particle strengthening technique in enhancing the alloy strength. The main enhancement is coming from the dispersion of Ag_3Sn IMC in tin matrix. Definitely, more Ag can be added to make the alloy modulus higher, but it will increase the possibility of Ag_3Sn platelets formation. Besides that, the high Ag-bearing Pb-free alloy has lower drop impact resistance. High Ag content in the alloy system will add more burden to the material cost too. Therefore, other approach is needed.

Solid solution strengthening is one of the options in strengthening the Pb-free alloy. In fact, solid solution strengthening had been widely applied during the SnPb soldering era where the Pb is added into Sn to improve the alloy characteristics. Somehow, this technique was not

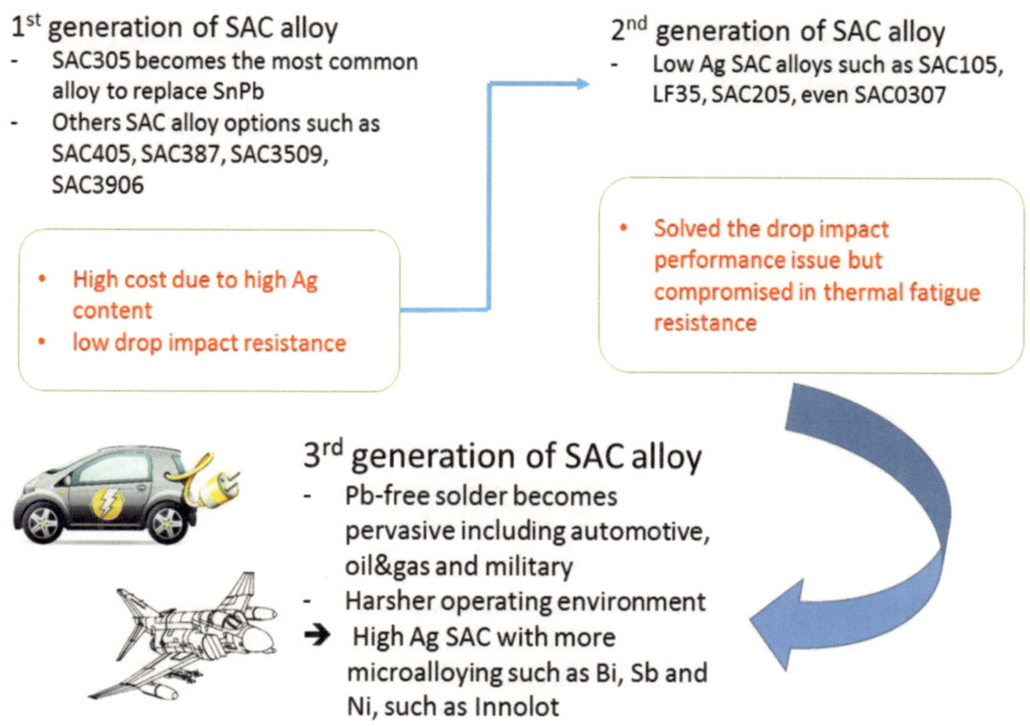

Figure 7. The evolution of Pb-free alloys from first generation to third generation, high reliability alloy for harsh use condition.

common during the conversion from SnPb to Pb-free soldering. In Pb-free alloys, the Cu and Ag are added to tin, separately or together, to make the most widely used SAC alloys, which have almost no solubility in the β-tin matrix. The Cu and Ag appear in the microstructure only as the intermetallic compounds, Ag_3Sn and Cu_6Sn_5 as discussed earlier. Surely, Pb cannot be considered again as solute in solid solution strengthening because of the RoHS compliance. Other solutes should be considered to achieve this objective, for example, Bismuth, Bi; Indium, In; and Antimony, Sb. These three substances are the most common and possible candidates selected by the solder manufacturers and researchers in enhancing the strength of Sn-based Pb-free alloy. Bi and In can decrease the liquidus of Sn-based alloy but Sb can increase the liquidus. It depends on the requirement of end application. If a solder alloy with lower homologous temperature is needed, microalloying with Bi or In may not be appropriate. However, Sb is a banned or restricted used substance in some applications such as mobile and consumer industries. It has limited the use of this substance for microalloying to improve the solder strength.

In early period of Pb-free soldering adoption which was about 2006, there was a working group formed to develop a more robust Pb-free alloy for automotive industries. Since then, the industries understand the need of a stronger alloy for future auto electronics requirement. This working group members include Siemens, Bosch, Heraeus, Alpha Metals, Infineon, the Fraunhofer Institute, and etc. 90iSC was the outcome of this group's effort. The 90iSC is a high Ag SAC alloy with addition of Bi, Sb, and Ni. It is a very complex system in which the characteristics and reliability cannot be fully understood and assessed. Solder manufacturers such as Nihon Superior has adopted a simpler approach by just micro-alloying Bi into SnCuNi system in order to achieve a stronger alloy. It is SN100CV®. Without the Ag addition into SN100CV®, the worry of alloy degradation due to Ag_3Sn coarsening can be lifted. Many users may still be skeptical on the Pb-free alloy without Ag addition. They are worrying about the strength of this Ag-free Pb-free solder. In fact, the solid solution strengthening with Bi addition is even more effective in increasing the tensile strength of the alloy. For example, in the SN100CV®, the 1.5% Bi microalloying can significantly increase the tensile strength of the SnCuNi alloy and make it even higher strength when comparing to SAC305 which containing 3% Ag. **Figure 8** is the comparison of tensile strength of three alloys, which are SnCuNi (SN100C®), SnCuNiBi (SN100CV®), and SnAgCu (SAC305).

Besides the as-cast comparison, the three different alloys shown in **Figure 8** have been subjected to thermal aging at 150°C for 500 h. After the thermal aging, tensile strength data were collected. As expected, there is significant degradation in strength for SAC305 sample due to the Ag_3Sn coarsening. The drop in tensile strength for both SnCuNi and SnCuNiBi is minimum. **Figure 9** summarizes the comparison data after thermal aging.

However, choosing a right ratio of Bi into SnCuNi system is challenging. If too little of Bi added, the effect of solid solution strengthening is negligible. But if too much of Bi added, precipitation of Bi out of β tin will occur. **Figure 10** shows the tensile strength of different alloys before and after thermal aging at 150°C for 500 h.

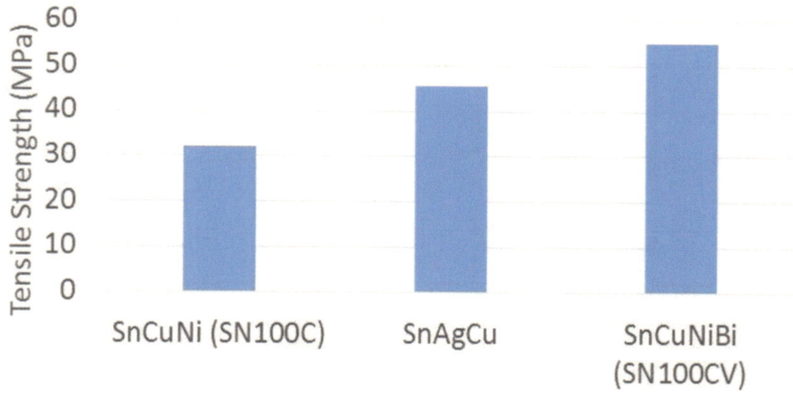

Figure 8. Tensile strength comparison for three alloys in as-cast condition.

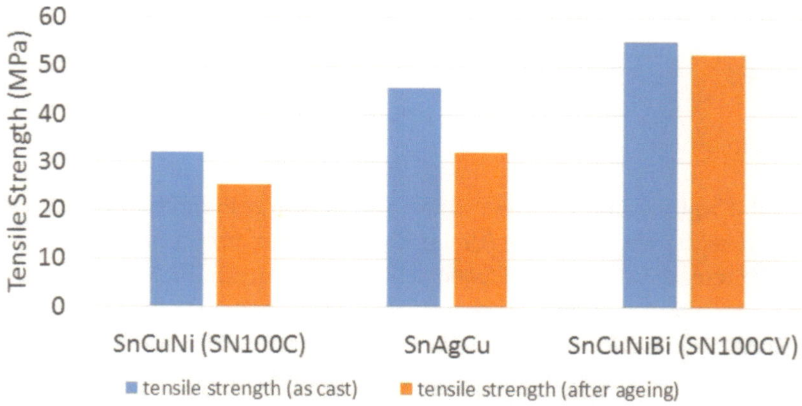

Figure 9. Tensile strength comparison for three alloys in as-cast and after thermal aging.

3.2. High reliability solder alloy for die attach

Solder usage is not limited to the interconnects between electronic components and PCB at the PCBA process. Within the electronic components such as integrated circuit (IC), solder has been used as die attach material especially in the device where there is a need of high thermal and electrical connectivity between the die backside and lead frame or substrate. This kind of IC includes the high power devices and high speed switches such as Power MOSFET, insulated-gate bipolar transistor (IGBT), high power diode and transistor, rectifier, and inverter. Due to high operating temperature of such IC, which is usually above 260°C, the industries are still exempted from RoHS compliance, and they are still using the high Pb solders such as Pb5Sn and Pb5Sn2.5Ag in die attach process. It is because the conventional Pb-free Sn-based alloys have melting point lower than 260°C. The RoHS committee allows another 5-year extension for this exemption in 2016 because of technical limitation with current technology. So far, there is still no straightforward drop-in solution in replacing the high Pb solder for such application.

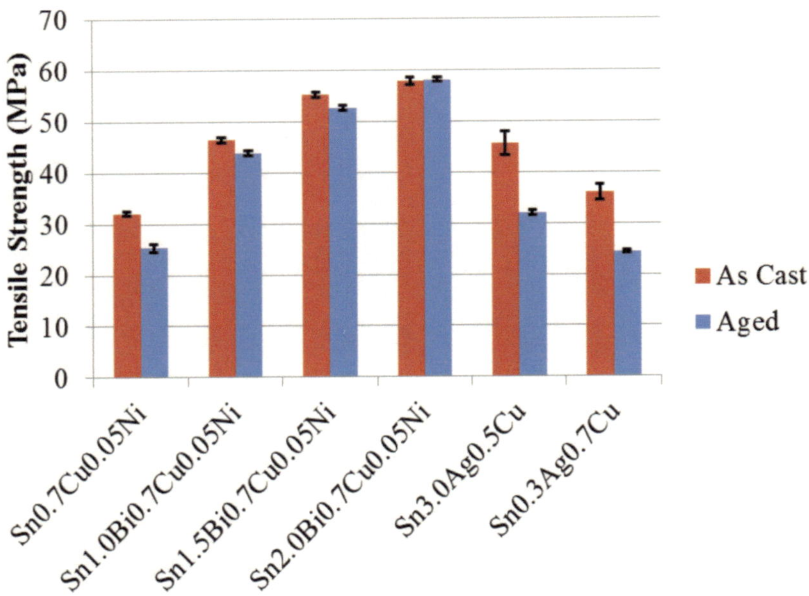

Figure 10. Tensile strength of test alloys as-cast and after thermal aging at 150°C for 500 h.

Since most of the Sn-based Pb-free alloys have solidus lower than 260°C, other Pb-free alloys such as gold tin (AuSn), bismuth silver (BiAg), and zinc aluminum (ZnAl) are being considered as candidates to replace the high Pb solder, which is the current high-temperature die attach material. However, there are some limitations in these alloys which the manufacturers need to overcome before the mass adoption is possible. The limitations include high cost (AuSn), low thermal conductivity, poor wettability (BiAg), and low workability due to brittleness (ZnAl). Indium Corporation has innovatively improved the characteristics of BiAg to make it more user-friendly product for the die attach process. To improve the wettability of BiAg, this company has introduced a solder paste consists of two types of solder powder. The first type is the major powder which is the BiAg and the minority powder which is the additive powder. The additive powder has lower melting point than BiAg and better wettability on common soldering pad. The additive powder can be the familiar SnBi eutectic, SAC alloy, or SnAg alloy. In the reflow process, the additive powder will melt first and react with soldering pad then follows by the melting of majority powder which is the BiAg. It is an irreversible process because the minority powder will be fully consumed or reacting with the majority powder during the reflow process. The joint formed using this solder paste will have similar properties as the BiAg, which is the majority part of this solder paste. More studies are required to assess the suitability of this solder paste in replacing the high Pb alloy. The solidus of BiAg is still lower than the high Pb solder. Therefore, the homologous temperature is different for these two solders.

Besides solder alloys, the industries are also working actively to develop sintering material to replace high Pb solder. The driving force to develop sintering material is not only to replace the existing high Pb solder but also to prepare a solution for future application. Based on the power IC roadmap, the IC operating temperature can go up to as high as 600°C especially for the silicon carbide (SiC) die. With such high operating temperature, it will be a challenge for conventional solder alloy. Sintering is an atomic diffusion process. Heat and pressure can be applied to accelerate this diffusion process. The atoms in the powder particles diffuse across the boundaries of the particles, fusing the particles together and creating one solid piece (bulk material). In other words, a lower process temperature than the material melting point is required to complete the sintering process. In sintering process, there is no melting involved. The joint formed after sintering will have the characteristics similar to the bulk material. Since Ag is the most common major ingredient for sintering material, the joint formed with Ag sintering material will re-melt only when the temperature exceeds the Ag melting point which is 961°C. This is the main attraction of Ag sintering material to be used in this application. In fact, the industries had explored Ag sintering material in 1980s. But, it still required high sintering temperature and pressure to complete the process during that period of time. It made the acceptance of this material low. With the recent development of nano Ag particle, the sintering process can be completed within an acceptable sintering temperature range. Nihon Superior has developed a Ag sintering paste, Alconano®, which has demonstrated good quality of silver joint at die attach layer with sintering temperature as low as 200°C. This is only possible because of the nano size Ag particles which have high surface energy. The challenge of manufacturing Ag sintering paste is not only in producing the Ag particles, but also in the development of right passivation system to keep the particles in good condition until the sintering process starts. Alconano® has utilized alcohol as the passivation system, which forms alkoxides with silver atoms on the surface of the nanoparticle. The advantage of these chemicals in this application is that the oxygen-silver bond, which is strong enough to stabilize the nanoparticle during manufacturing processes and subsequent storage and handling, is weak enough that it can be broken at a relatively low temperature to expose the active surface of the nanoparticle so that it can bond to adjacent particles. Another advantage of using alcohol as the passivation system is it leaves no harmful residue behind after the sintering process. The residue is free from sulfur and nitrogen compounds that can interfere with the performance of the sintered silver and contribute to corrosion problems in service. The challenge of mass adoption of Ag sintering paste does not limit to paste manufacturing only. Application of this paste at die attach process is also challenging. Many engineering works are required in developing an appropriate set of process parameter. Sintered Ag joint is a porous layer. In fact, this porosity is necessary to make the Ag be a reasonable die attach joint. Bulk Ag joint without any porosity could be too rigid to support the IC architecture. In such case, die crack will be the most prominent defect in the field. Getting a right properties of this Ag joint should be a joint effort between the user and supplier. One of the wonder of this Ag sintering paste is the porosity can be adjusted via sintering process. By varying the pressure and temperature during sintering, different range of porosity can be achieved. **Figure 11** shows an example of Ag sintered joint of Alconano® with porosity less than 10%. This joints were formed with pressurized sintering parameters: 10MPa, 300°C and 5 min sintering time. The consistency of porosity percentage is satisfying. Besides satisfying the processibility, the reliability of joint made of this Ag sintering material must be on par or even better than the performance of high Pb solder. There are two main characteristics of such die attach material, which the users are emphasizing. They are thermal conductivity and electrical conductivity of the joint. Based on a recent industry wide survey on Pb-free high-temperature die attach material, thermal conductivity stands out to be the most required property for this material [7]. It is crucial to keep high thermal conductivity throughout the entire product life. In some cases, the joint has degraded thermal conductivity due to the thermal fatigue crack within the joint especially in hot and cycling atmosphere. Therefore, in developing the Alconano®, Nihon Superior has compared the performance of Alconano® versus the high Pb solder and conventional SAC solder after exposure to thermal cycling reliability testing, −40°C/+200°C with dwell time of 30 min. The comparison results are shown in **Figure 12**.

Pb-free liquid-phase diffusion bonding (LPDB) material is another emerging material, which can potentially be used as high-temperature Pb-free die attach material. The bonding process of this LPDB is very similar to the modified BiAg paste mentioned above. Similar to the modified BiAg paste, the LPDB paste consists of two types of powder, high melting point and low

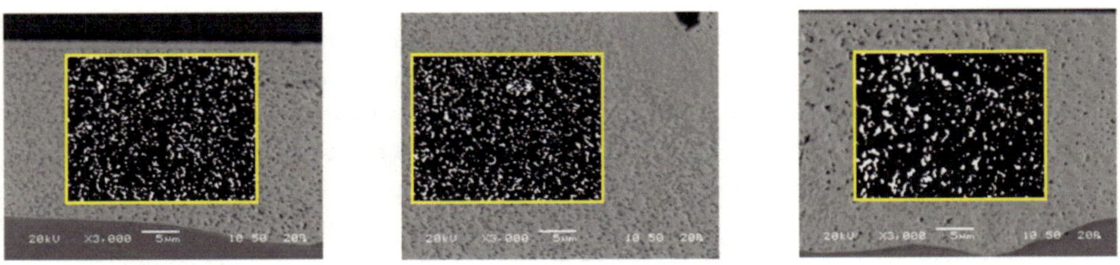

Porous rate: 7.67% | Porous rate: 7.05% | Porous rate: 8.79%

Figure 11. Porosity check after pressurized sintering on Ag sintering joint.

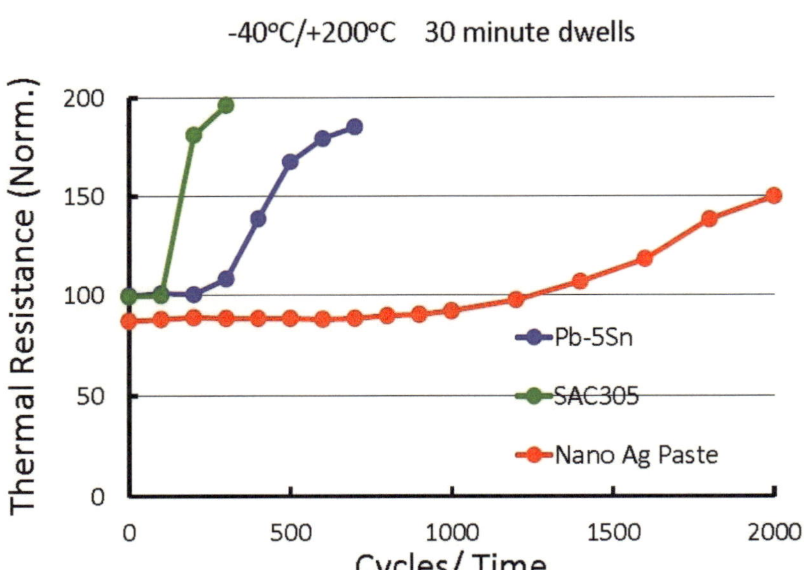

Figure 12. Change in thermal resistance of high temperature die attach material in a function of thermal cycling. 10×10 mm silicon die on Al_2O_3 substrate with Cu plate heat sink was used as test vehicle. The sintering condition was 300°C, 40MPa pressure assisted and 3 min sintering duration.

http://dx.doi.org/10.5772/intechopen.69553

melting point. Unlike the modified BiAg paste where both types of powder will melt during the bonding process, the high melting point powder of LPDB will never melt during the bonding process but only the low melting point powder will melt and react with soldering pads and simultaneously react with the high melting powder to form a new IMC. During the bonding process, homogenization of the molten occurs. IMC will be formed because of the reaction between the molten low melting point powder and high melting point powder. And, this newly formed IMC has much higher melting point than the low melting point powder. Therefore, the molten will start to solidify even before the cooling starts due to the change of melting point. After the bonding process, the joint will only remelt at a temperature higher than the bonding temperature. This is why this material is also called transient liquid-phase diffusion bonding material. Nihon Superior Co., Ltd. has participated a project run by Ames lab from Iowa State University in developing this LPDB material. Ames lab is mixing the high melting point Cu-10Ni powder into the commercial SN100C® fine powder to make this LPDB material. The bonding process will transform all SN100C® powder into high melting point $(Cu, Ni)_6Sn_5$ IMC which will only remelt at 525°C [8]. This is a very important characteristic to make it a potential high temperature Pb-free die attach material. Moreover, the Ni addition into Cu_6Sn_5 has significantly improved the properties of this IMC. It has made it a more robust joint because the Ni has inhibited the polymorphic transformation of the allotropic Cu_6Sn_5 as discussed earlier in this chapter. According to Choquette and Iver [8], the Ni addition into Cu_6Sn_5 should improve the ductility and strength of the joint as well. This LPDB material can be a potential drop-in solution in replacing the high Pb solder because it can complete the bonding at conventional reflow temperature (240–260°C), but it will only re-melt at 525°C.

4. Conclusion

Due to RoHS compliance, electronic industries have stopped adopting the stable and familiar SnPb-6337 solder, which was used as electrical interconnect for decades and migrated to Pb-free solder. There are many developments of Pb-free solder with the objective to fulfill the manufacturability and reliability expectation from the users. This kind of development has flooded the market with many alloy compositions and increased the difficulty for user to select a right Pb-free alloy, which can meet their expectations. Since 2006, generally, there are three generations of Pb-free solder being introduced into market. The first generation of Pb-free solder is the conventional SAC alloy with high Ag bearing ranging from 3.0 to 4.0%. Then, the market is moving into low SAC alloy due to the surge of Ag price in early 2010s and the poor drop impact resistance of these conventional SAC alloys. Recently, due to the aggressive growth of auto electronics and avionics, a new set of requirement in term of solder joint reliability is defined against the Pb-free solder. This is the third generation of Pb-free alloy, which is an alloy system catering for long thermal fatigue life and robust in drop impact resistance. On the other hand, there are still many IC and electronic components still adopting high Pb solder (soft solder) as die attach material. The industries are actively looking for Pb-free alternatives to replace it. Alloy systems such as AuSn, BiAg, ZnAl, Ag sintering material, and LPDB material are the potential candidates on the list.

Acknowledgements

This publication is fully funded by Nihon Superior Co., Ltd. The author is grateful to Nihon Superior Research and Development, R&D department for their support in data collection and analysis. The author would also like to express his special gratitude to Dr Tetsuro Nishimura, Mr Takatoshi Nishimura, and Mr Keith Sweatman for their help and guidance on this publication.

Author details

Wayne Ng Chee Weng

Address all correspondence to: wayne@nihonsuperior.co.jp

Nihon Superior Co. Ltd., Osaka, Japan

References

[1] The European Parliament and of the Council. Restriction of Hazardous Substances Directive 2002/95/EC (RoHS 1), EU. 2003. Available from: http://eur-lex.europa.eu/LexUriServ/LexUriServ.do?uri=OJ:L:2003:037:0019:0023:EN:PDF

[2] Deloitte. Trends and Outlook of the Auto Electronics Industry, China. 2013. Available from: https://www2.deloitte.com/cn/en/pages/manufacturing/articles/trends-and-outlook-of-auto-electronics-industry.html

[3] Suganuma K, editor. Lead-Free Soldering in Electronics. USA: Marcel Dekker, Inc.; 2004

[4] Nogita K, McDonald SD, Tsukamoto H, Read J, Suenaga S, Nishimura T. Inhibiting cracking of interfacial Cu_6Sn_5 by Ni additions to Sn-based lead free solders. Transactions of the Japan Institute of Electronics Packaging. 2009;**2**(1):46-54

[5] Pandher R, Healey R. Reliability of Pb-free solder alloys in demanding BGA and CSP applications. In: Electronic Components and Technology Conference; 27-30 May 2008; USA. 2008

[6] Coyle R, et al. A collaborative industrial consortia program for characterizing thermal fatigue reliability of third generation Pb-free alloys. In: SMTA International 2016; 25-29 Sep. 2016; Rosemont, IL, USA. 2016

[7] Pei LS, Pan B, Zhang H, Ng W, Wu B, Siow KS, Sabne S, Tsuriya M. High-temperature Pb-free die attach material project phase 1: Survey result. In: International Conference on Electronics Packaging (ICEP2017); 2017; Japan. 2017

[8] Choquette SM, Anderson IE. Advances in the research of a Sn/Cu-Ni composite solder paste for high temperature use. In: SMTA International; 25-29 Sep. 2016; Rosemont, IL, USA. 2016